BarracudaGraphs

Flightline Series 1

de Havilland Sea Vixen FAW 1 and 2

British Naval All Weather Jet Fighter

An in-depth photographic study by
Roy Sutherland

Langhorne Publishing House, Inc.
www.langhornepublishing.com

ISBN-10: 0 9842591 0 4
ISBN-13: 978 0 9842591 0 6

First Edition

10 9 8 7 6 5 4 3 2 1
09 10 11 12 13 14 15 16 17 18

Langhorne Publishing House, Inc.
San Francisco, CA

Contact us at
info@langhornepublishing.com

Visit our website at:
www.langhornepublishing.com

Acknowledgements

The production of this book started almost 20 years ago, even though I did not realize that it had until a few months ago. I have been collecting information on the Sea Vixen for all that time, and my file on the Sea Vixen is stuffed with everything I could lay my hands on. Along the way, I have been helped by many people, and I'd like to take this opportunity to thank everyone. Being almost 20 years in the making, my biggest fear is that some people's help will go unacknowledged. That, and bears. I will do my best, but please know that if you helped and do not see your name here, you have my sincerest apologies.

First off, I'd like to thank my son Cooper, who has had to put up with me working long hours over the last few months on this book. I'd like to thank my business partner Chapman Chan, for keeping me focused and working, doing such a great job on the design and layout of the book, his knowledge of computers and magical voodoo stuff like FTP sites, and for his friendship. It was Chapman who pushed me to launch this new series. Mike Grzebien and I spent a miserable cold and rainy November day at the Newark Air Museum, shooting and measuring their Sea Vixen. Thanks Mike!

Thanks to my friends Chris Bucholtz for writing the history section of this book, and John Heck, who designed the BarracudaGraphs logo and is my webmaster extraordinaire.

The heart of this book is the photos and illustrations. Originally, the book was going to consist of my photos only, but I soon realized that I was woefully short on images. Mark Robson came to the rescue with his vast collection of closeup detail photos. Keith Bunyan spent many hours photographing and scanning his negatives. Many more gaps needed filling, and a number of guys came to the rescue, some at the last minute. The following is a complete list of image contributors to this book. All photos are the Copyrighted property of the photographers who took the pictures (or captured the images) and are used with their kind permission.

Roy Sutherland	Chris Hughes (UK)	Robin Powell	Peter Woolley
Mark Robson	Peter Steehouwer*	John Heck	Chris Clifford
Robin Walker	Roland Bogush	Keith Bunyan	

A number of other friends helped me make sense of this incredibly complex aircraft. Thanks to Robin Powell, Mike Grzebien, Martin Hale, Derek Pennington, Lee Coll, Drewe Manton, and Mark Robson. Thanks to the Newark Air Museum and the Mosquito Air Museum for giving me access to their Sea Vixens many years ago.

* Check out more of Peter Steehouwer's aviation photography on his website: www.steehouwer.com.

Flightline Series 1
de Havilland Sea Vixen FAW 1 and 2

The de Havilland Sea Vixen represents the zenith of British twin boom jet fighter design. This all-weather carrier borne aircraft was the first British fighter to be designed without guns. Armed with rockets and missiles, the Sea Vixen, with its integrated weapons systems, transitioned the Royal navy into the era of modern fighter aircraft. Take an in-depth look at the inner secrets of this sinister looking aircraft and travel back to the days before microprocessors and miniaturization. A tangle of wires, cables, pushrods, bellcranks and boxes await you.

Roland Bogush photo.

CONTENTS

Foreword

Frog Sea Vixen

From as far back as I can recall, I've always loved aircraft, especially British Aircraft. My dad, who was born and raised outside of London, England, was building models of Spitfires Hurricanes and Messerschmitts in 1939, when the kits consisted of a set of plans and some rough cut wooden components. As soon as he was old enough, he joined the RAF, and by the end of the war, was flying Lancasters with 1654 HCU. When the war ended, there was no flying to be had. He finished his service and left the RAF.

In the mid-1960s, he was back to building models, this time in plastic kit form. I was enthralled. As soon as I was old enough, I started building with him. "What the heck does this have to do with the Sea Vixen?" you may ask. Don't get your knickers in a twist... I'm getting to it. My first exposure to de Havilland's twin engined, twin boom fighter was to come from a cardboard box up in the attic that contained some of my dad's unbuilt models. In there was a Frog Sea Vixen. The Frog kits at the time had some terrific, evocative box art. I saw it and I was hooked.

Some 20 years ago, maybe more, I was in England visiting family. My cousin Paul took me to the Mosquito Air Museum. I was shocked to see a Sea Vixen FAW 2 there. After talking with the staff for a while about this aircraft, they opened it up for me and allowed me to climb in the cockpit and shoot photos until my wallet started complaining. If you recall, back then, shooting images cost money! So, that day with my Uncle Paul was the real beginning of this book, even though I did not realize it until recently.

I'm a detail junkie. For me, there is never enough. Some modelers are enthralled with the history and the exploits of daring young men in their flying machines. Not me. I love the machines. Stick a camera in some dark orifice of an aircraft and reveal what is in there, and you have my attention. I am always on the hunt for books that show close up images of cockpits and gearbays. There are detail type monographs out there, and some do a better job than others of covering the aircraft in question, but I usually find myself wishing that they had just stuck the camera back .... there... and maybe another one over... there.

So, with some pushing and prodding from my good friend Chapman, BarracudaGraphs and the Flightline Series was born. The Sea Vixen was a natural first title for me, as it scratches a 20 year itch. I think you will find that the coverage is pretty comprehensive. Did we get everything? Probably not, but there is a goldmine of new information in these pages. I wanted to get photos of a fully restored Martin Baker Mk 4DSA seat outside of the cockpit, but time ran out. Space precluded the addition of the buddy pack refuelling pod. This IS the first book in the series, and as such has been a trial by fire. The next book will continue the format, but there will be improvements as well.

A quick note about the photos. I make no apologies for the quality of the images. The vast majority were taken with 35mm SLR Cameras on color print film, most of them at least 15 years ago. Negatives had to be scanned and cleaned up in Photoshop. Just cleaning up the dust and scratches eats up a lot of time. To compound the problems, the weather on the days we were shooting always seemed to reflect the designated mission of the Sea Vixen; Fighter (All Weather). That is to say, the weather ranged from cold and overcast to cold and raining. Shooting in low light leads to images of lesser quality, being thin and lacking in contrast and "snap", as well as messing with the colors. I was able to correct a lot of the problems in Photoshop, but there are a few images throughout the book that are not of great quality, but are included because they fill in a missing part of the picture.

Future titles will be comprised of digital images wherever possible. This will result in a higher level of definition and detail, better color balancing, and will virtually eliminate out of focus and poorly exposed images. Our aim is to bring you the most complete coverage possible and at a reasonable price. The focus will be on the equipment. There are plenty of sources out there for in-depth histories. We value your opinions. We encourage you to contact us and let us know what you think. What did we do right? What could we do better? What other features would you like to see from this series? What subjects would you like to see? Thanks for purchasing this first title in the series. Our goal is to provide the modeler and aviation fan with a range of high quality books that will fill the gaps in your reference library.

Happy modeling!

Roy Sutherland
Author and Master Modelist*

* yes, we know its not a real word!

© Crown Copyright.

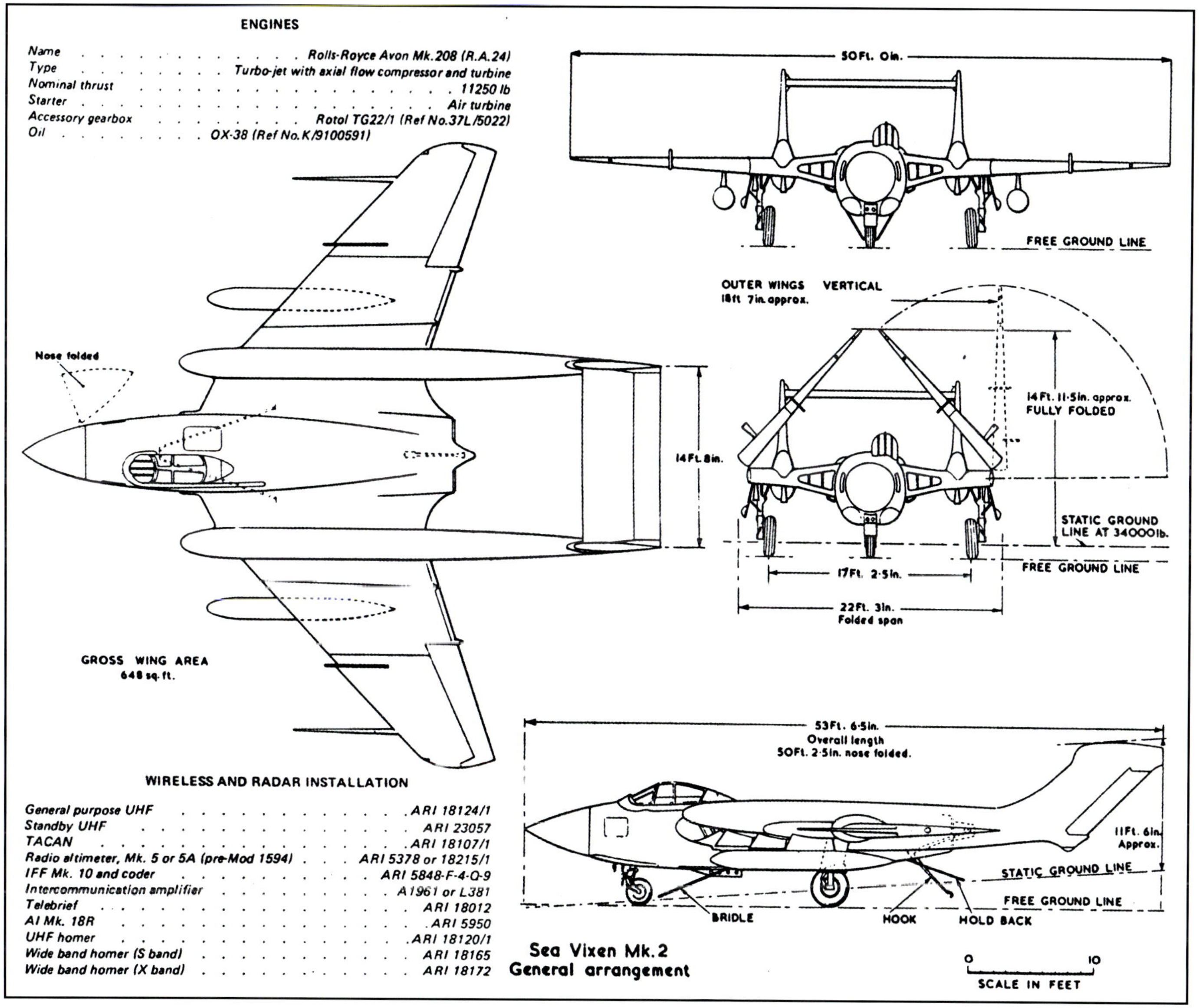

History of the De Havilland Sea Vixen by Chris Bucholtz

Sea Vixen FAW 1 XJ492 rotating during takeoff. Marked as VL-715 of 766 Squadron. Yeovilton 7/7/1963. *Robin Walker photo.*

The Sea Vixen represented the apex of the twin-boom fighter design, a lineage pioneered by the Lockheed P-38, Northrop P-61 Black Widow and Saab J.21A. The arrangement made the leap into the jet age with the Saab J.21R, but it first really achieved its potential as a jet planform with the de Havilland Vampire. The Vampire was built in great numbers – 3,268 – and was the first-generation jet for many air forces. It also set the pattern for de Havilland fighter designs for the next 15 years.

Next came the Venom, which expanded on the Vampire design and gave it greater flexibility. A total of 1,431 Venoms, Sea Venoms and French-built Aquilons were manufactured and again the fighter saw widespread use, continuing in use until its 1983 retirement from the Swiss Air Force.

The Vampire and Venom were great aircraft, but they were also products of their times. The jets were straight-winged, blunt-nosed and flew with engines (the Goblin in the Vampire, the Ghost in the Venom) that represented the early stages of advanced jet engine technology. They had another thing in common: both aircraft went to sea in carrier-specific versions. This aspect of de Havilland's fighter designs would play a key role in the development of the Sea Vixen.

Vixen Beginnings

Even as the Sea Vampire was entering service, de Havilland and the admiralty were discussing the possibility of a more modern design with two engines, a feature especially desirable in a naval aircraft in an era of jet engines of questionable reliability. The added safety margin a second engine provided safeguarded pilots, and it also promised an increase in performance. The twin-boom layout was proven to be a winner aboard carriers; it allowed for easy storage, access to the engines for maintenance was straightforward, and visibility over the nose was good. It would also permit the two engines to be placed side-by-side in the fuselage, ensuring that an engine-out situation would not result in an uncontrollable asymmetric thrust problem for pilots.

At the time of these discussions, which began in 1946, data from captured German research was beginning to have an impact on British, American and Soviet designers. As a result of this data and data generated in the de Havilland DH. 108 project, the new aircraft was envisioned with a wing swept at 40-degree angle. The plane would be an all-weather fighter armed with four 30mm cannon placed below the nose, firing below the radar dome.

Even at this early stage, the two-man crew was projected to be positioned in a unique arrangement, with the pilot under a canopy offset to port and the observer tucked away in the right side of the fuselage slightly behind the pilot, below a flush-mounted hatch. To avoid a total lack of spatial awareness, small windows were provide for the observer, but the dim signals on early airborne intercept (AI) radar meant that ambient light had to be kept to a minimum.

In January 1947, Naval Specification No. 40/46 and RAF Specification F.44/46 were issued and called for very similar aircraft: a night interceptor that embodied the characteristics of the plane first discussed in 1946. The RAF specification mandated an aircraft that could achieve 525 knots at 25,000 feet, and could climb to its service ceiling of 45,000 feet in less than 10 minutes. It needed to be able to engage enemy aircraft at a maximum altitude of 40,000 feet, and its minimum endurance (including a 15-minute climb to 25,000 feet and 15 minutes of combat) was set at two hours.

De Havilland's response was the DH. 110; its competition was the Gloster GA.5, which would evolve into the Javelin. The DH.110 was a cantilever mid-wing aircraft of all-metal stressed-skin construction, with the trademark de Havilland twin boom layout. The wings were three-spar structures, with integrally-machined skins; Fowler flaps were fitted to the trailing edge.

Cutbacks, Changed Minds and a Crash

In April, 1949, the Ministry of Supply ordered seven land-based night fighters and two long-range fighter prototypes for the RAF and a further two night fighters and two long-range strike fighters for the Fleet Air Arm. However, in November of that year the Royal Navy selected the less advanced therefore less risky Sea Venom as a replacement for its de Havilland Sea Hornets, and the RAF cut its orders back to two DH.110s. These were assembled at Hetfield and the first of these planes, WG236, took to the air for the first time on September 26, 1951, with John Cunningham at the controls. The aircraft featured a nose that was very similar to that of the Sea Venom, giving it a bulbous appearance compared to what was to come in the Sea Vixen.

Testing of WG236 and WG240, the second aircraft, progressed for a year before the Farnborough Air Show in 1952, when the project suffered perhaps its worst setback. John Derry and observer Tony Richards were putting WG236 through its paces, flying fast and low toward the crowd, when the plane broke up in mid-air. Both aviators died when the fuselage hit the airfield; one of the engines continued and whirled into the audience, killing 29 and injuring many more. WG240 was grounded; investigation showed the crash was caused by torsional failure of the

Sea Vixen FAW 1 XJ514 in the shadow of a Blackburn Beverly. Assigned to AHU Brawdy. Finningley 19/9/1966. *Robin Walker photo.*

Top. Sea Vixen FAW 1 XJ517 taxying out in the rain. Marked as VL-719 of 766 Squadron. Yeovilton 29/6/1963. *Robin Walker photo.*

Bottom. Sea Vixen FAW 1 XJ521 parked on the ramp. Marked as R-243 of 890 Squadron. Yeovilton 27/6/1964. *Robin Walker photo.*

wing cause by a combination of high speed and rate of roll, which peeled back the leading edge of the wing. This area was reinforced and the tail was redesigned on the surviving prototype, but at about this time Gloster's Javelin was ordered by the RAF and the DH.110 project seemed doomed.

Luckily, the Admiralty was still hoping for an all-weather fighter, and issued a new requirement in 1952. A swept wing Venom was initially considered before the DH.110 regained favor; the revised aircraft would feature a higher thrust Avon engine, increased fuel capacity and modifications to improve its shipboard qualities. The 30mm cannon were retained initially, but the new aircraft would be based around the Blue Jay missile, which would soon be renamed the Firestreak.

Touch-and-go landings by WG240 aboard HMS Albion showed the DH.110 could indeed go to sea. The Admiralty ordered a third aircraft, the DH.110 Mk. 20X, which carried an arrestor hook and long-stroke carrier-compatible landing gear, but it lacked radar and folding wings. Its first flight in June 1955 was followed by carrier suitability trials on HMS Ark Royal in April the next year, but by then the Admiralty had ordered the new fighter into production.

Sea Vixens for the Fleet

Initial orders of the Sea Vixen FAW Mk. 1 called for 78 aircraft, including 21 pre-production aircraft that would be used to develop the aircraft and its weapons system. As changes were made, the pre-production planes could be retrofitted and new build aircraft could incorporate the changes; this way, the preproduction aircraft could themselves become squadron aircraft should they survive the development process.

The Sea Vixen was redesigned to specification N.139P; by the time all was said and done, it represented an 80 percent redesign of the DH.110. Hydraulically-operated folding wings, catapult hooks, and nose wheel steering were added to improve carrier handling qualities; the pilot's canopy was redesigned; a large ventral airbrake was added, and the engine placement was shifted slightly within the fuselage. The booms were reset with a few degrees of upward crank to improve clearance on rotation, and a pointed radome replaced the bulbous Venom-style nose. Among the most significant changes was the deletion of the 30mm cannon; these were replaced by by two trays of 14 two-inch rockets carried forward and to either side of the nose wheel well. This made the Sea Vixen the first British fighter to fly without guns.

The first production aircraft, XJ474, was pushed out of the hangar at

Sea Vixen FAW 1 XJ567 carrying Fireflash missiles. Marked as VL-718 of 766 Squadron. Yeovilton 29/6/1963. *Robin Walker photo.*

Christchurch for the first time in February 1957 and took to the air for the first time on March 20. It served in handling trials, deck acceptance work and spin trials before spending the rest of its career at Royal Air Establishment Bedford. The second and third aircraft (XJ475 and XJ476) had similar careers in the development of the Red Top missile and never saw squadron service. The earliest-build aircraft to reach squadron service was XJ477, which was used for weapons trials around the rocket armament before being passed to 766 Squadron in August 1962.

The FAW.1 Into Squadron Service

Long before that, however, the Sea Vixen had gone into an operational role. First to fly the aircraft in the fighter role was 700Y Squadron, the Sea Vixen Intensive Flying Trials Unit, at Yeovilton, which formed on November 3, 1958. This unit was charged with carrying out an evaluation of the type; when finished, in June 1959, it disbanded and became the core of 892 Squadron, which embarked on HMS Ark Royal with 12 aircraft. In October, four of these planes became 892 A Flight, and became the Firestreak IFTU while the remaining eight stayed with the ship's air group. 892 B Flight set sail on Ark Royal again in March 1960, and when 892 A Flight completed trials in August, it rejoined the squadron for the remainder of the cruise.

In late September, the squadron disembarked to Yeovilton and then in October redeployed to HMS Victorious, which made a long Far East cruise starting in January 1961. The change of decks was a calculated move, as the Admiralty wanted to give as many ships' crews the experience of working with the jet as it could, and so in May 1962 the squadron shifted to HMS Hermes, then to HMS Centaur in December 1963. It was during this cruise that the Sea Vixen first saw combat. After flying air defense and surface search missions during the undeclared war between Indonesia and Malaysia over the future of the island of Borneo, the Centaur was on its way home when its air wing was tasked with flying strike missions during the Aden Crisis against National

History of the De Havilland Sea Vixen continued

Left. Sea Vixen FAW 1 XJ576 with "RN Fighter School" above code. Marked as VL-711 of 766 Squadron. Yeovilton 29/6/1963. *Robin Walker photo.*

Bottom. Sea Vixen FAW 1 XN651 with running fox art on the tail. Marked as C-461 of 893 Squadron. Yeovilton 29/6/1963. *Robin Walker photo.*

Liberation Front guerillas in the Radfan. The Centaur then served to show the flag during unrest is Dar-es-Salaam in East Africa before returning home, where 892 Squadron disembarked to re-equip with the Sea Vixen FAW. 2.

While 892 was at sea, the first aircraft were delivered to 766 Squadron, the navy's all-weather fighter training unit. It stood up at Yeovilton as an all Sea Vixen unit on October 24, 1960 after phasing out its Sea Venom FAW.21s. Earlier that year, the next Sea Vixen fighter squadron, 890 Squadron, was formed and in July it went to sea aboard HMS Hermes for a week of work-ups before departing on a Far East cruise later in the year. In 1961, the squadron crossed decks to the HMS Ark Royal and made four more Far East cruises before it disbanded on October 7, 1966.

The third fighter unit to fly the Sea Vixen FAW.1 was 893 Squadron, which was established on September 9, 1960. In early 1961, the squadron took its six Sea Vixens aboard HMS Ark Royal for 10 days of familiarization and cold-weather training in the Davis Straits; after completing this training, the squadron moved to HMS Centaur and an Indian Ocean cruise in April 1961. The ship was part of a British show of force in the Persian Gulf during the handover of Kuwait, and for the next three years 893's Sea Vixen FAW.1s flew from the vessel's decks. In 1964, the squadron transferred to HMS Victorious for another Far East cruise. When Victorious returned home, 893 was disbanded.

On February 1, 1961, 899 Squadron stood up at Yeovilton to serve as the type's headquarters squadron. This unit was tasked with training and operational procedures and weapons trials for the Sea Vixen FAW.1, and maintaining a pool of experienced Sea Vixen crews as a operational reserve. 899 Squadron also conducted the service trials of the Sea Vixen FAW.2 starting in early 1964, but once this was accomplished the squadron shed its headquarters squadron status and took its FAW.2s to sea as an operational squadron aboard HMS Eagle.

Ultimate Sea Vixen: the FAW.2

Even as the Sea Vixen FAW.1 entered service, a set of improvements was in the works. Initially, these were to be rather dramatic; one proposal was to replace the Avon with a less thirsty alternative, the Rolls Royce Spey. At the other end of the spectrum, it was suggested that afterburners be added to the engines and 250-gallon wingtip fuel tanks and an 850-gallon tank in an extended fuselage be fitted to offset the possible expenditure of fuel. Both proposals would have greatly increased the weight of the aircraft, so the proposed aircraft, the DH.110 Mk.21, would have used blown flaps to lower approach speed and improve low-speed handling characteristics. These improvements were deemed too ambitious, requiring extensive re-design of the entire aircraft, so a more modest set of improvements was made to create the Sea Vixen FAW.2.

Most evident among these changes was the addition of saddle tanks on the booms, which now projected ahead of the leading edge of the wings. These tanks reduced speed by five miles per hour, but they extended the aircraft's flight time to three hours without refueling. That was the other endurance modification made to the FAW.2 – it was fitted with an in-flight refueling probe. Additionally, the introduction of "buddy" refueling pods to the fleet in 1960 meant that Sea Vixens could operate in the tanker role as well.

Perhaps more important to the Sea Vixen's effectiveness was a change of missiles from the Firestreak to the Red Top. The Firestreak was a heat-seeking weapon, but its sensor head was lacking in sensitivity, meaning that it had to be fired from within a narrow cone immediately behind its target if it was to guide properly. The Red Top's seeker head was much more sensitive, to the extent that it could detect heat sources such as friction on a target's leading edges. The ventral rocket packs were deleted, and a redesigned clear-vision canopy was added after FAW.2 production had begun.

The FAW.2 program stipulated the construction of some new-build aircraft and the modification of many more FAW.1s to the new standard. In 1961, two new FAW.1s, XN684 and XN685, went from the Christchurch production line straight to Hatfield for conversion into the FAW.2 prototypes. The modified aircraft first flew on June 1, 1962, and both prototypes were sent to No. 13 Joint Service Trials Unit for development of the Red Top in 1964. Two years later, these prototype aircraft were fully converted to current FAW.2 standard and went to 893 Squadron aboard HMS Hermes.

XP919 was the first new-build Sea Vixen FAW.2, making its first flight on March 8 1963. In all, 29 new-build machines were constructed at Hawarden. At the same time, conversion work to the FAW.1 fleet proceeded apace, and by the time the final conversion had been completed 67 aircraft had been modified.

The FAW. 2 Goes to Sea

The FAW.2 entered squadron service with 899 Squadron, at the time the Sea Vixen headquarters squadron, and after the final FAW.1 left the unit it became an operational squadron, going aboard HMS Eagle in December 1964 and remaining part of the ship's air wing until Eagle was decommissioned in 1972.

Top. Sea Vixen FAW 1 XN696 with spiral painted refueling probe. Marked as VL-488 of 899 Squadron. Yeovilton 30/6/1962. *Robin Walker photo.*

Bottom. Sea Vixen FAW 2 XJ576 touches down with full flaps. Marked as V-244 of 893 Squadron. Yeovilton 9/9/1967. *Robin Walker photo.*

Top. Sea Vixen FAW 2 XJ580 with Redtops on a seemingly rare sunny day. Marked as VL-724 of 766 Squadron. Yeovilton 9/9/1967. *Robin Walker photo.*

Bottom. Sea Vixen FAW 2 XN658 taxying out on a foggy morning. Marked as 750 of 890 Squadron. Yeovilton 6/9/1969. *Robin Walker photo.*

899 Squadron had the distinction of being the final Sea Vixen front-line unit.

766 Squadron, the training unit, was next to get the FAW.2, starting in July 1965. Over the next three years, the squadron had the trying task of maintaining and training for both marks of aircraft, and it wasn't until 1968 that it finally became an all-FAW.2 unit. By this time, the phase-out of the Sea Vixen was beginning, and 766 Squadron itself was disbanded in 1970. 890 Squadron assumed its training duties.

Next up to receive the new fighter was 893 Squadron, which reformed at Yeovilton with 11 aircraft on November 4, 1965. The squadron went to sea aboard HMS Victorious for a Far East cruise in 1966-67, but while the ship was in dock for maintenance the vessel suffered a fire which damaged her beyond economical repair. With its previous home carrier on her way to the breakers, 893 Squadron shifted to HMS Hermes in May 1968 for a cruise to the Mediterranean and east of Suez, followed by a final Far East cruise in 1970. Following this deployment, Hermes was sent to the yards for conversion to a commando carrier, and the book closed on 893 Squadron's Sea Vixen career. It was disestablished on July 14, 1970.

The penultimate front line user of the FAW.2 was 892 Squadron, who were fully re-equipped with 12 aircraft in December 1965. In 1966, the unit went aboard HMS Hermes for a Mediterranean and Far East cruise before returning to Yeovilton. With time for the squadron growing short, and the squadron still landlocked, the men formed an aerobatic team called "Simon's Sircus," after commanding officer LCDR Simon Idiens, and made the rounds of U.K. airshows. The end of summer 1968 was followed by the end of 892 Squadron, which was disestablished on October 4.

When 899 Squadron went to sea, it left the Sea Vixen without a headquarters squadron. That was remedied by the establishment of 890 Squadron, which stood up on August 14, 1967, equipped with a mixed bag of FAW.1s and FAW.2s. It retired the older planes in July 1969, but the squadron's planes were withdrawn and passed to the Royal Naval Air Doctrine School (ADS). The squadron had been working with the school to train direction officers, but the decision was made to allocate the aircraft to the school itself. After remaining grounded for a short time, 890 Squadron was provided with aircraft from the recently-deactivated 766 Squadron. The FAW.2s were welcome additions, since the headquarters squadron now had a training role, but by 1971 it was decided that no new crews needed to be trained, and on August 6, 1971, 899 Squadron itself was disbanded.

The End of the Road

The removal of the Sea Vixen from carrier operations – and indeed the U.K.'s wholesale exit from the fixed-wing carrier business – did not mean the immediate extinction of the de Havilland fighter. The ADS' Air Direction Training Unit (ADTU) was still operating the plane, even through a series of mergers between units to create the Fleet Requirements and Air Direction Training Unit (FRADTU), then the shortened Fleet Requirements and Air Direction Unit (FRADU). The A&AEE at Boscombe Down operated three Sea Vixens, and in these training and research roles, the Sea Vixen could be seen flying through 1974.

As the aircraft fell into end-of-the-line roles instructional airframes for battle damage repair, destructive testing subjects, firefighting practice subjects – one role kept the Sea Vixen flying: that of a drone aircraft. The Gloster Meteors which had filled that role were, by 1972, becoming scarce, so the possibility of the Sea Vixen U.3 drone aircraft was contemplated by the Royal Air Establishment Farnborough. All equipment unnecessary for operations would be removed from the aircraft – specifically, armament and radar – and the aircraft would be equipped for control from a ground station or another aircraft while retaining its ability to be flown by an on-board pilot. The observer's "coal hole" made an idea place to fit the equipment for recording miss distances and other weapons metrics, and the aircraft had the performance needed to present a realistic simulation for the weapons systems then in development.

From 1972, ample Sea Vixens were passed to RAE Farnborough for storage and potential conversion, but drone work progressed very slowly. In fact, only one such aircraft had been completed, XN657, before the project was transferred from Tarrant Rushton to Hurn in 1980. No official specification was issued for what became the Sea Vixen D.3, and the program suffered from a chronic lack of funding, finally ending entirely in 1983. By that time, two more drones, XP924 and XS577, had been converted. All three aircraft were delivered to RAE Llanbedr; XN657 was used infrequently and was scrapped in 1986, but the other two aircraft flew with some frequency. In 1991, XS577 was grounded and became a parts source for XP924, which flew on until the end of the year. This seemingly ill-starred program preserved the single airframe airworthy in the 21st Century, as XP924 was restored and still makes the rounds of the U.K.'s airshows.

Sea Vixen FAW 2 XN685 carrying two 36x 2 inch rocket pods. Marked as VL-703 of 890 Squadron. Yeovilton 17/7/1971. *Robin Walker photo.*

Fuselage - Forward Fuselage

Standing back to give a good idea of the true shape of the forward fuselage and radome, an area so often misrepresented in model kit form. The tricycle wheeled tank is a travel pod with towbar attached. *Roy Sutherland photo.*

Another view of the pilots canopy and the prominent rain dispersal duct fairing. Note the sliding hood is a different late frameless style, without the prominent white sealing strips. The large hatch provides access to the radar equipment bay, the smaller door, to the back of the instrument panel. *Roy Sutherland photo.*

Looking forward at the starboard side of the nose. The large hatch is also for access to the radar equipment bay. Ahead of that is the complex radome hinge mechanism. The bulged cross section of the retractable microcell launcher is visible in this image. *Mark Robson photo.*

The revamped observers hatch and it bulged fairing are not the only changes for this crew station. The observer's side window has also been replaced with a new design. The dashed yellow outline is an axe chop marking. *Mark Robson photo.*

The port side of the nose . Forward of the Radar access hatch is the latch for opening the radome, and the stencilling detailing its operation. The NACA duct provides much needed cooling air to the jam-packed radar equipment bay. *Mark Robson photo.*

This shot of the lower port fuselage shows a number of interesting details, including three small access panels and a small brass probe that might be a temperature sensor or a probe for the altimeter. Behind that is a venturi. Its purpose is also unknown. *Mark Robson photo.*

The starboard microcell rocket launcher shown highlighted for clarity. This extendable pod housed 14 folding fin unguided two inch rockets per pod. The rockets could be fired singly, rippled, or as a salvo. *Mark Robson photo.*

Another view of the Observer's ladder. "Observer" is a somewhat anachronistic title. With that tiny side window, I'm not sure how much observing could be accomplished. Radar Operator is better, but Systems Manager may be closer to the truth. *Roy Sutherland photo.*

The Observer's ladder in position. The top step hooks into a slot in the fuselage and the ladder is locked into position with the turn of the handle on the lower mounting plate. *Roy Sutherland photo.*

Fuselage - Upper Fuselage

Left. Looking back along the upper fuselage of a Sea Vixen FAW 2. The red "do not walk" markings have been worn and faded to near white over the years of sitting outside. The engine bay panels are prominent in this view. Note the twin VHF antennas. Outboard of these are the port and starboard fuel filler hatches. *Mark Robson photo.*

Right. This view shows the forward portion of the upper fuselage, and the rear view of the canopy and bulged late-style observers hatch. Note missing UHF antennas and the partially open port fuel filler door. The engine and accessory door latches need some work on this aircraft. They fit well on an operational Sea Vixen. *Roy Sutherland photo.*

In the foreground, the Ram Air Turbine, or RAT, has been deployed. In the event of complete hydraulic system failure, this turbine could provide enough hydraulic pressure to glide the aircraft in for a landing. In the background, note the early observers hatch, easily identified by the small window. *Roy Sutherland photo.*

This view shows the RAT to good advantage, as well as the rear of the upper fuselage. The turbine retracted into the top of the hookwell. With the hook down and the RAT deployed, you can see daylight through the open turbine doors. *Roy Sutherland photo.*

Left. With the upper surface engine doors removed, the upper portion of the main firewall could be removed and the engines lifted out by crane for servicing and replacement. The NACA ducts provided fresh air and cooling to the engine bay. *Roy Sutherland photo.*

Right. This view shows the engine accessory bay doors. The back of the bulged Observer's door fairing is removable to allow the access hatches to be removed. The rear end of the rain dispersal duct fairing is seen. This duct splits off from the port intake de-icing hot air pipe, and runs externally forward to the windscreen, covered by the prominent fairing. *Mark Robson photo.*

The black squares are screens that have been stuck down over the various intakes, and are not present on aircraft in flight. The view shows the upper surface of the fuselage and inner wing. The location of the wing fold jury strut in the upper wing can be seen. *Mark Robson photo.*

Curves everywhere! Call me crazy, but I think the Sea Vixen is a beautiful aircraft. The well worn and very flat finish of this FAW 2 is actually a boon to the modeler, as it shows every detail very well. This general view is full of useful information. *Mark Robson photo.*

The many intakes and vents on the upper fuselage are detailed here. The 4 NACA ducts supply air to cool and vent the engine bay. The D-shaped grill is a bleed air escape duct. *Roy Sutherland photo.*

Fuselage - Lower Fuselage

Left. Underside shots of the Sea Vixen are a challenge, as it sits pretty low to the ground! Here is a general view of the undersurfaces of the center fuselage area. The large engine access panels are prominent. In the foreground is the closed mainwheel door. *Mark Robson photo.*

Right. A similar shot from the starboard side. This area is a veritable forest of access panels, drains, vents, bleed doors, and doors for the connection of ground equipment such as generators, gas charging, and testing carts. *Chris Hughes photo.*

Right. This intake serves a dual purpose. The outer inlet provides cooling air to the generators, and the inner one feeds the air conditioning system. Starboard intake shown. Port intake is identical. *Mark Robson photo.*

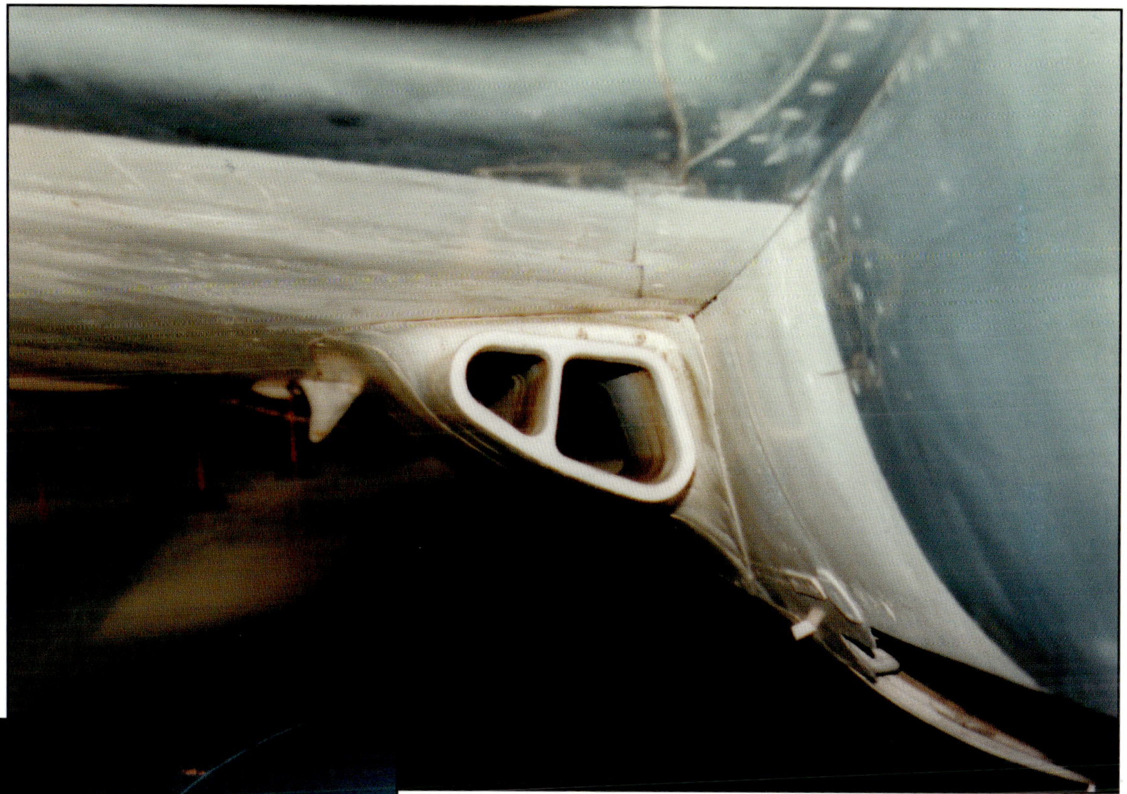

Left. This odd little exhaust is the wheelcase breather vent. It is one of a pair located at the trailing edge of both wingroots. A small detail often overlooked on the Sea Vixen. *Mark Robson photo*

Left. Looking aft along the underside of the port wing at the wingfold joint. Note the wingfold fairing and the missing underwing pylons, showing the mounting points. Again, the somewhat worn finish of this FAW 2 is a plus for us, as it shows the panel details in stark relief. *Mark Robson photo.*

Bottom. A view of the underside area forward of the airbrake shows the three hatch doors that allow access to the maze of equipment under the cockpit floor. The antenna 's specific purpose, like others on this aircraft, is unknown. *Mark Robson photo.*

Left. The leading edge of an FAW 1's starboard wing from below. The two light stores pylons are not fitted, and the mounting recesses have fairing panels installed. Note the inboard one is from a different aircraft, as the paint demarcation does not match. *Mark Robson photo.*

Right. The starboard catapult strop hook is a very solid casting. It needs to be, to take the strain of repeated catapult launches. Getting a 42,000 lb. aircraft from zero to flying speed in a few seconds involves a lot of force and acceleration! *Chris Hughes photo.*

Top Left. This almost perfect bottom view of the last flying Sea Vixen is terrific for making sense of all the details photos in this section. Clearly seen are the catapult hook, light store and droptank pylons, wingfold fairing, wing fence and dogtooth, as well as the partially deployed flaps. *©Peter Steehouwer photo.*

Below. The same photo again, but this time focusing on the forward fuselage. Note the airbrake and associated access doors, the nosegear, and the clearest view yet of the retractable microcell rocket launchers. The true shape of the radome is obvious here. Note the non standard location of the forward blade antenna. *©Peter Steehouwer photo.*

Bottom Left. A different area of the same photo reveals the underside of the fuselage in the same light. Note the closed airbrake, the ram air intakes for the a/c, the closed mainwheel doors, the lowered hook, and the mass of engine and accessory access panels. *©Peter Steehouwer photo.*

Lower Fuselage

Lower Fuselage

Fuselage - Lower Fuselage continued

Fuselage - Airbrake

Left. View from port side looking forward at the Sea Vixen's massive airbrake. Note that the hydraulic jack is disconnected from the door on this airframe, and the typical staining seen on the interior of the airbrake on in-service aircraft. *Roy Sutherland photo.*

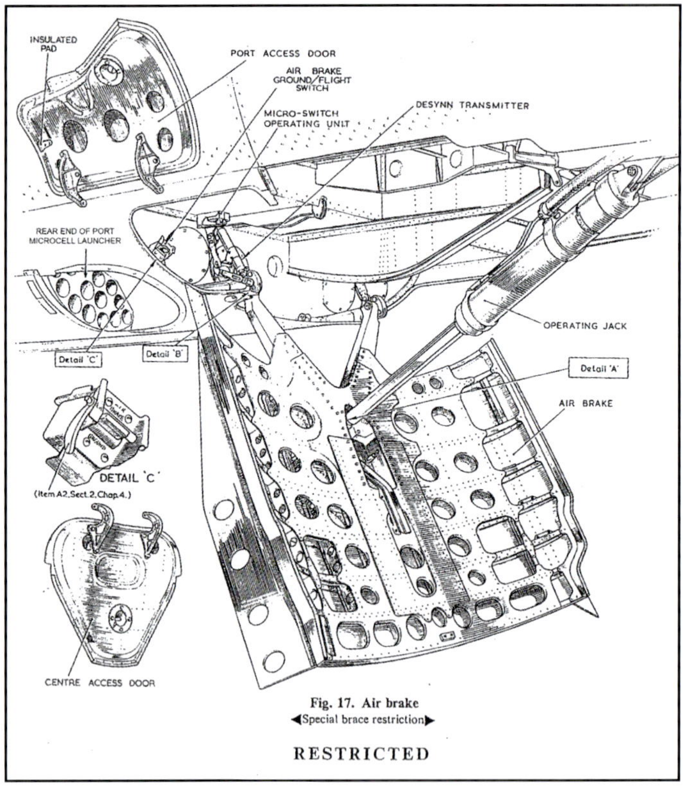

Right. Note how the operating jack attaches to the door, and the removed cover, showing the rear face of the port retractable 14 round Microcell unguided rocket launchers. This view is also useful, as it shows the roof and structural members of the bay, without all the equipment, wiring and plumbing installed. © *Crown Copyright.*

Left. The complex system of electronics necessary to operate a state-of-the-art Airborne Interception system in the 1960's is exposed in this shot of a Sea Vixen FAW 2 with the radome open. Today, all this equipment could probably be fit into one or two compact boxes thanks to the amazing advances in integrated circuitry in the last forty years. *Roy Sutherland photo.*

Right. The bottom of the radar equipment bay is filled by two large electronic components. On the left, looking like a large beer keg, is the servo unit, and on the right is the transmitter/receiver unit. *Roy Sutherland photo.*

Left. The fiberglass radome could be hydraulically folded by the pilot. This was necessary to allow the Sea Vixen to squeeze onto the deck elevators aboard fleet carriers. Note the numerous locks to secure the radome when closed, and the large radar dish scanner. *Roy Sutherland photo.*

Right. The scanner dish gimballing mechanism and the scanner dish are seen to good effect in this image. The scanner could be aimed by the observer, using the joystick mounted in his cockpit. Note the large manual locking latch on the right of the photo. *Roy Sutherland photo.*

Radome continued

Top Left. The scanner shows the limit of vertical travel. Note the substantial radome hinge and the power folding mechanism attached to an arm on the hinge. The front ring of the scanner is fiberglass, while the dished back is made of a perforated metal screen. *Roy Sutherland photo.*

Above. Arrayed across the top of the equipment bay are 3 large canisters. From left to right, these are: The wave form generator, the power pulse unit, and the radar modulator. A number of smaller components and a maze of electrical cables and hydraulic lines make this bay very busy and visually interesting. *Chris Hughes photo.*

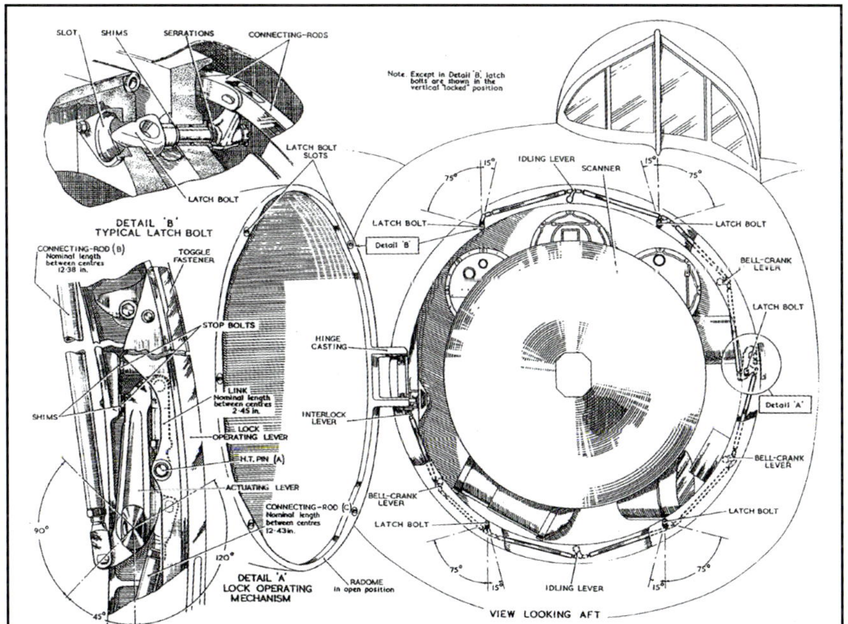

Fig.5 Radome lock mechanism
RESTRICTED

Left. The radome hinge and locking mechanism are detailed in this technical illustration. It also shows well the very distinctive offset pilots canopy of the Sea Vixen. The large number of parts necessary to simply lock the radome, gives an idea of just how incredibly complex a modern fighter aircraft is. Keeping a machine like this flying is a very expensive and labor intensive job. © Crown Copyright.

Radome continued

Cockpits - Pilot's Cockpit

Left. Looking into the pilot's cockpit. The main instrument panel and starboard console are shown. The large vertical panel on the pilot's right side is an electrical panel, with banks of switches for lights, radios, booster pumps, fuel management, armament and stores jettisoning. The cable bundle on the floor would be stowed in service. *Roy Sutherland photo.*

Right. A slightly different view . Note the sign on the canopy sill is non-standard. The red handle is the manual canopy winder in position (usually stowed below the sill). The aircraft is steered on the ground with the large vertical handwheel under the electrical panel, which turns the nosewheel. The right side of the main instrument panel is home to the engine rpm, jet pipe temperature, fuel and flow gauges. Above these is the autopilot panel. *Roy Sutherland photo.*

Left. The horizontal panel on the pilot's right side. At the top of this console is the UHF control panel. Below that is a panel for the Mk 17F oxygen regulator. To the right of that are the engine start controls. The blue rectangle is a warning panel, and it is flanked by fire warning lights. Note the rudder pedals and very worn floor. *Roy Sutherland photo.*

Right. The so -called "blind flying panel" is edged in white. Top row from left to right: mach meter, airspeed, attitude, and heading indicators. Bottom row: (missing) Red Top indicator, altimeter, slip (above), angle of attack (below), rate of climb and artificial horizon indicators. The row below that, L to R: air brake, homing, tailplane, flap, and fuel rate indicators. *Roy Sutherland photo.*

Above. The port console and throttle quadrant. The yellow and black striped handle is the hood jettison release. Below that is a gray handle that controls the wingfolding. Towards the back of the cockpit sill is a pair of gauges covered by a clear plastic shield. These are the pressure gauges for the "red" hydraulic system, which drives a motor that supplies power to the AI radar system. *Roy Sutherland photo.*

Right. The small upper panel, usually in shadow from the coaming. This contains the LABS controls, flight refueling controls and lights, de-icing switches. The blank plate is the old position of the missile readiness indicator on the FAW 1. The rocker switch on top of the control column is for tailplane trim. *Mark Robson photo.*

Top Left. The three colored gauges display pressure for the different hydraulic systems. Blue supplies power for the flying controls and the autopilot. Yellow is a backup that duplicates the Blue system. The Green system operates the flaps, undercarriage, arrestor hook, brakes, wing folding and their latch pins, and nosewheel steering. Aft of them are the twin wooden throttle handles. *Mark Robson photo.*

Bottom Left. At the far right are pair of grey levers. These are the LP fuel cocks. Below that is another grey lever that controls the flaps. The yellow and black marked lever aft of that is the tailhook control. Behind that is the cabin pressure lever. The four grey knobs (two are in shadow) are lighting dimmer switches, and the brown dome is the cabin heating control. *Mark Robson photo.*

Left. Back to the starboard console. Aft of the blue panel and the warning lights is the TACAN controller. This is a military navigation system that is similar to the civilian VOR nav. system. Behind that is a small joystick for the autopilot. Note the lower pull ring and the blue leg lines on the ejection seat. *Roy Sutherland photo.*

Right. A view of the pilot's port sidewall from the observers position. This angle shows details of the upper sidewalls usually unseen when shooting from the cockpit.

The dome with the lever is a friction adjuster for the throttle. The oval yellow and black shielded pushbutton is the clear wings jettison switch! The placard warns that flaps, gear, and hook must not be deployed above 220 knots. *Mark Robson photo.*

Left. An overhead view looking down into the cockpit. The rails of the canopy track, the inflatable pressure seal, the windscreen and the sliding hood are all clearly shown. The Martin Baker Mk 4DSA was a good seat, but the extreme acceleration from the explosive ejection charge left some crew with spinal compression injuries; A situation remedied with the change over to rocket motors. *Roy Sutherland photo.*

Right. The starboard console from a different angle. Behind the joystick is a filler plate where the radio altimeter used to be. Aft of them are the two white-handled wing fold and wing lock controls. The yellow and black lever deploys the ram air turbine in emergency situations. The last knob is the flap interlock override. *Mark Robson photo.*

Left. Ejection seats packed a lot of functions into a small package, and grew more complex with each new model. The Mk. 4DSA seat was quite advanced for the time, but it was not a zero-zero seat; that is, a seat capable of being fired at zero altitude and at zero forward speed. *Roy Sutherland photo.*

Right. A tangle of harnesses and connections greets the aircrew who straps one on. A four part seat harness, a five part parachute harness, leg straps, r/t lead, oxygen lead... all that detail makes for an interesting and colorful focal point in a Sea Vixen's cockpit. Note the black and yellow D-ring. When pulled sharply down, a canvas screen covers the pilot's face, and starts the ejection sequence. *Roy Sutherland photo.*

Left. Looking down through the canopy at the top of the seat. The headbox contains the drogue chute, which is pulled out by a steel pin fired a few seconds after the seat leaves the aircraft. The drogue chute then inflates and pulls the main parachute out of the seatpack. The Sea Vixen's two seats were similar but different. *Roy Sutherland photo.*

Right. The shelf behind the seat, like just about every other space on the Sea Vixen, filled up with junction boxes, electrical wiring and assorted plumbing. Note the hot air duct running up the center of the rear pressure bulkhead. *Mark Robson photo.*

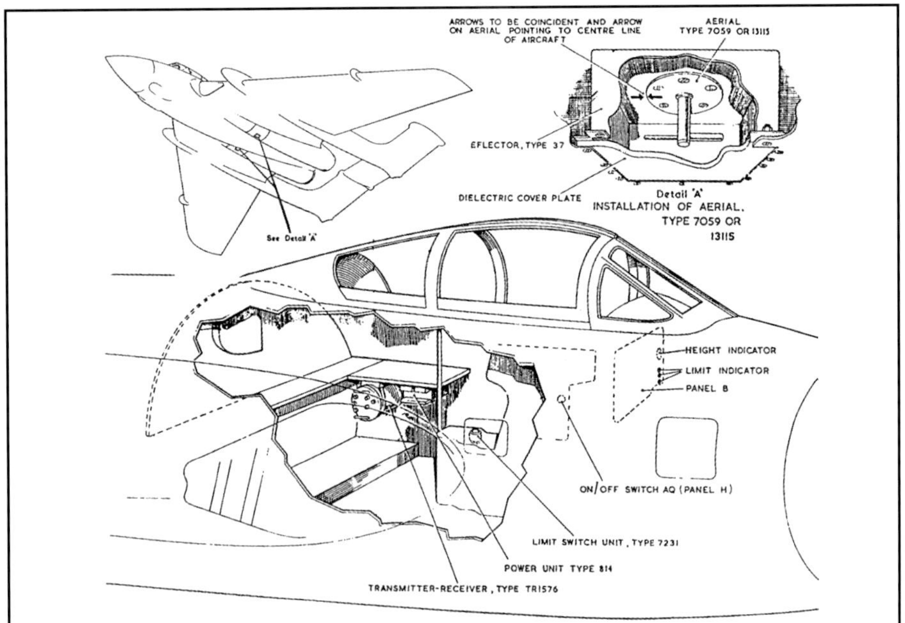

Fig. 1 A.R.I.5378 (Radio Altimeter Mk.5) installation

RESTRICTED

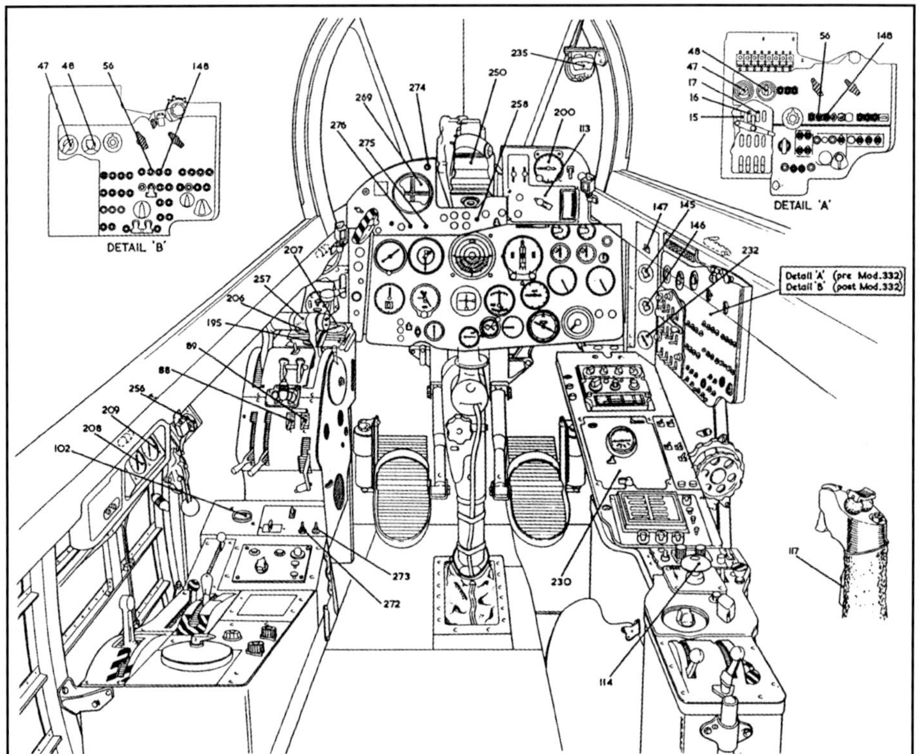

Fig. 8 Instruments and controls, pilot's cockpit

RESTRICTED

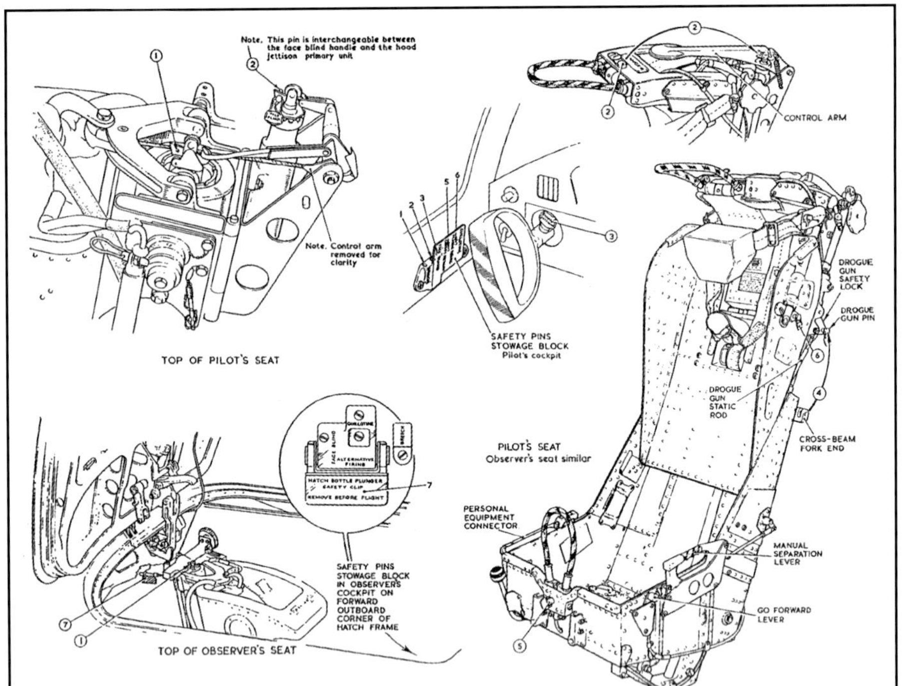

Fig. 1. Ejection seat safety devices

RESTRICTED

Top Left. An interesting diagram answers a question about some of the unidentified panels on the underside of the fuselage. This drawing also clearly shows the layout of the cockpit floor and rear bulkhead and how they relate to the airframe. I love this kind of reference. © *Crown Copyright.*

Top. This cockpit diagram is helpful in make sense of all the confusing shapes and colors in the accompanying photos. It also shows two different layouts for the electrical panel. © *Crown Copyright.*

Bottom Left. A great collection of drawings shows the basic structure of the Mk 4DSA ejection seat from a number of different angles. Note the pin storage block on the port canopy sill and the green metallic personal equipment connector. © *Crown Copyright.*

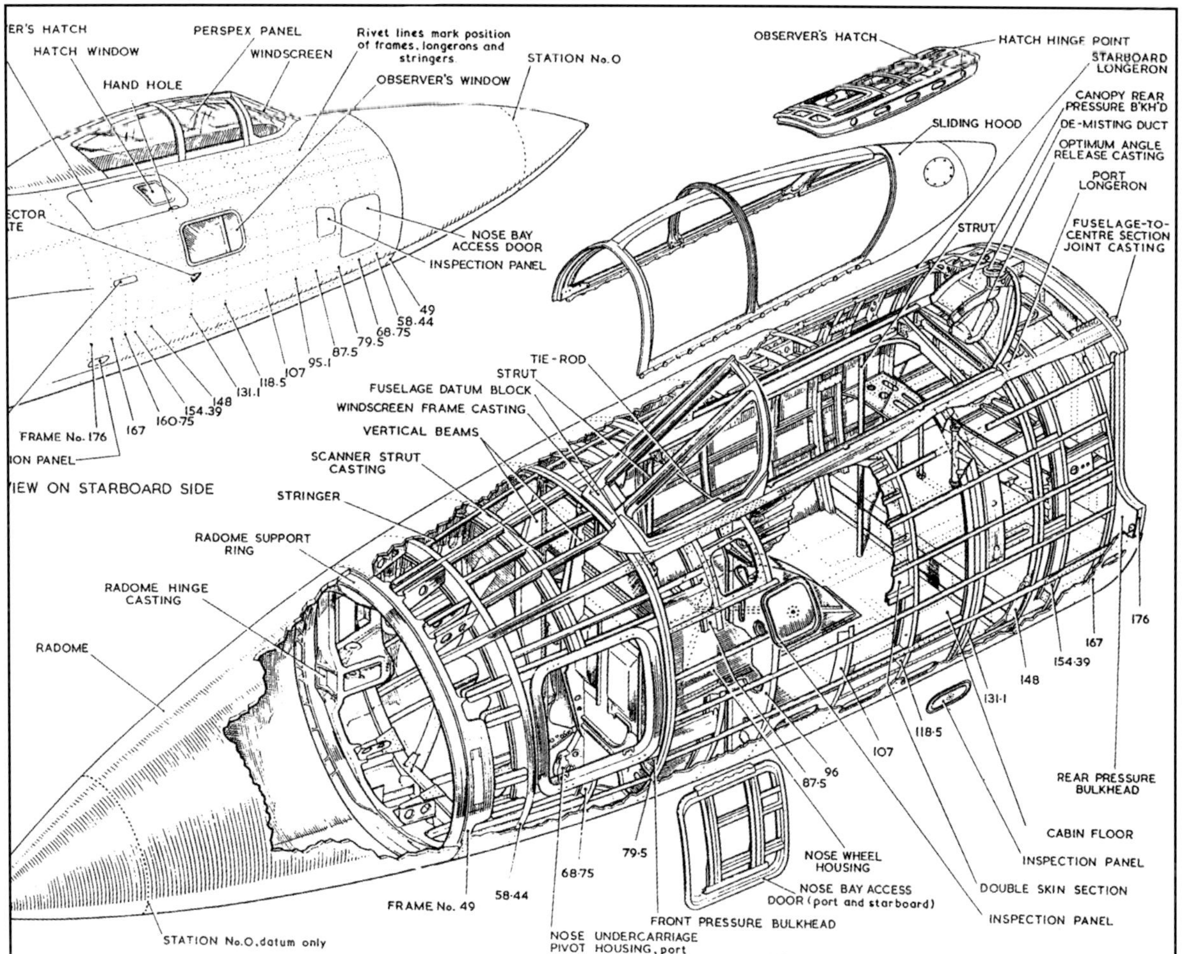

Fig. 1. Fuselage construction
SEA VIXEN FAW Mk.I AIRCRAFT

Left. I can't tell you how much I love illustrations like this. Not only is it an impressive piece of draftsmanship, but it is an amazing source of information, helping the engineer or modeler understand so much about the structure of the Sea Vixen. © *Crown Copyright.*

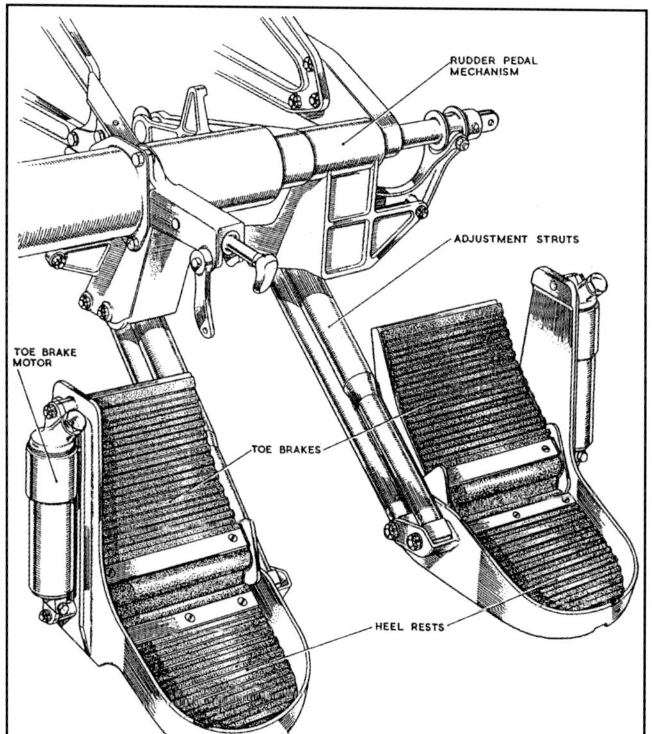

Fig.17 Rudder pedals
RESTRICTED

Right. The rudder pedal assembly, incorporating toe brakes. This useful illustration shows the detail that is hard to capture on film in the somewhat small and cluttered cockpit. © *Crown Copyright.*

Left. This view clearly shows the canopy seal support strip, with its black rubber inflatable seal. Also visible are the four hot air escape ducts for the stbd. pane of the windscreen. Just visible are two of the four ducts for the port side. *Roy Sutherland photo.*

Right. Head on view of late-style windscreen shows well the offset pilots cockpit. Note the late-style rain dispersal fairing at the base, with its raised channel that directs away boundary layer air and rain running back from the nose, and the rolled over top lip on the starboard half of the fairing that directed more rain away. *Mark Robson photo.*

Left. A great view looking down inside the rain dispersal fairing showing the hot air duct that replaced the windshield wiper assembly. The center beam of the windscreen houses a hot air duct that directs air between the spaced dual pane windscreen panels for de-icing and de-misting. *Mark Robson photo.*

Right. Not great weather for climbing around on airplanes! This general view shows the sliding portion of the pilot's canopy in the closed position. Note the rear portion of the hood is hinged. The rear cap can be folded up on top of the hood to allow removal of the fuselage accessory bay panels. *Roy Sutherland photo.*

The sliding hood viewed from the starboard side. Note the one piece perspex canopy, wide white pressurization sealing strip and many screws that seals the clear acrylic canopy to the sealing strip. The early style sliding hood did not have this prominent feature. *Mark Robson photo.*

The rain dispersal duct fairing wraps around the port side of the canopy and runs back along the top of the fuselage. This houses an external hot air pipe that runs forward from the upper fuselage, just forward of the engines. *Mark Robson photo.*

Note the late style sliding hood with the white sealing strips was also retrofitted to FAW 1s. The sliding hood is jettisonable for ejection. Seat fires one second after hood is blown. *Mark Robson photo.*

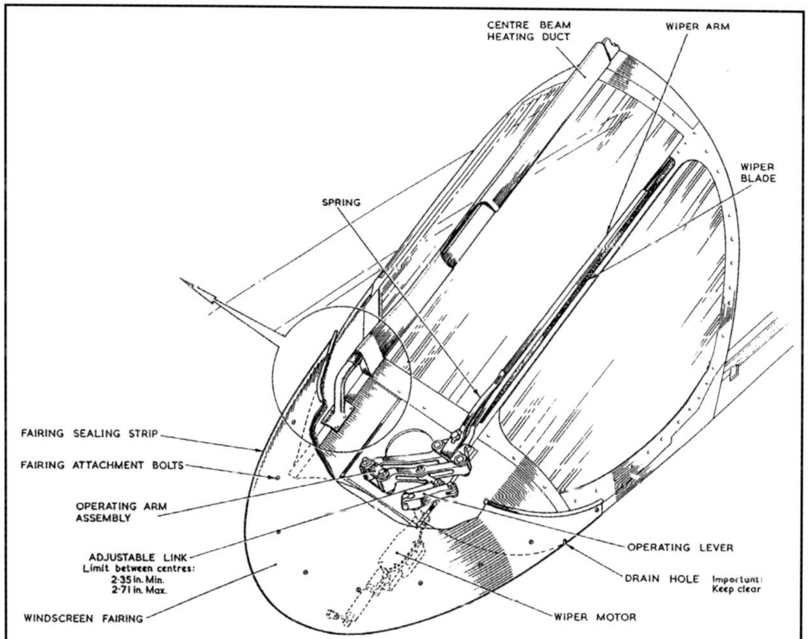

Fig.7 Windscreen wiper and heating ducts (post mod. 346)

RESTRICTED

No Sea Vixens photographed were fitted with the early style windscreen and small wiper fairing. This drawing shows to good effect the details of the early style unit as seen on FAW 1s before being replaced by the later model. © *Crown Copyright.*

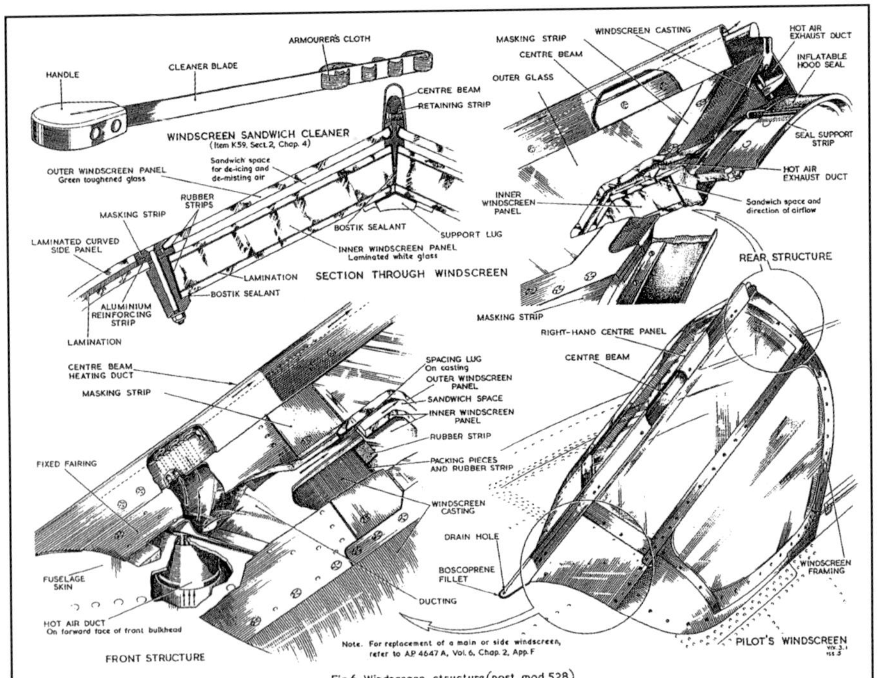

Fig.6 Windscreen structure(post mod.528)

RESTRICTED

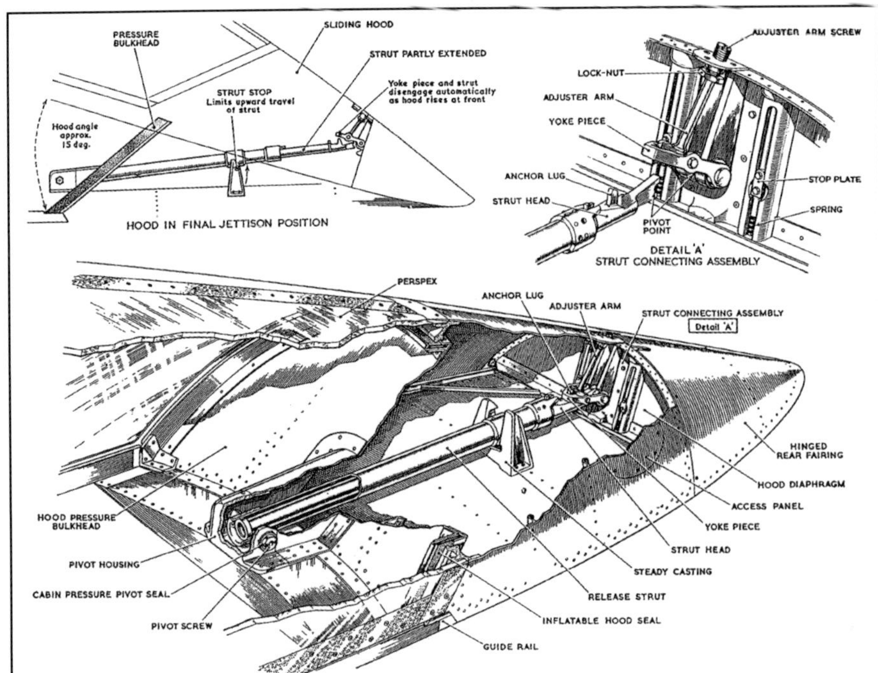

Fig. 7. Optimum angle release strut

RESTRICTED

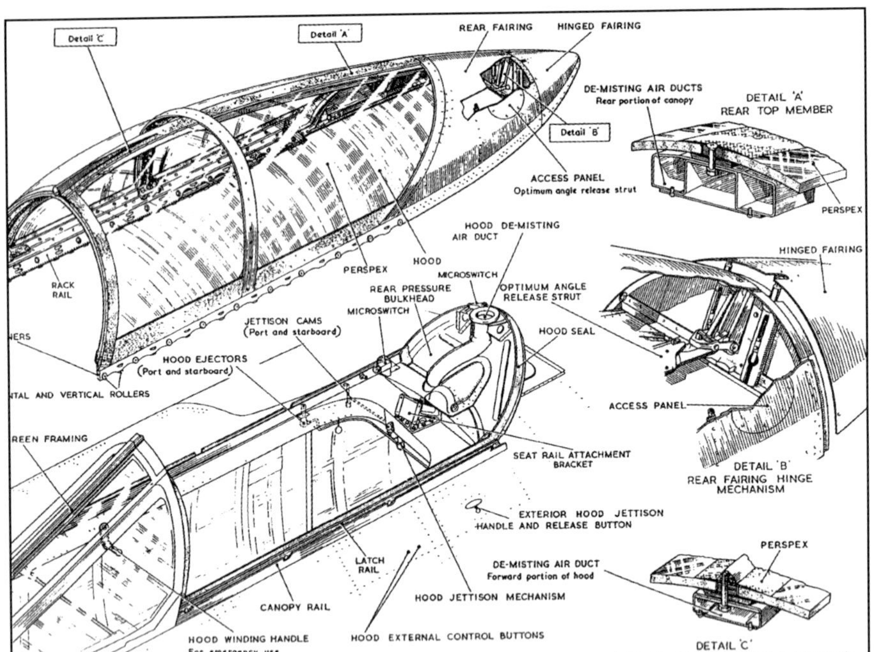

Fig.9 Canopy sliding hood
◄ Mirror deleted: microswitches added ►
RESTRICTED
SEA VIXEN FAW Mk.1 AIRCRAFT

Top Left. The complex structure of the late-style windscreen is detailed in this document. The rain dispersal duct and fairing are not shown. Note the spaced windscreen panes for demisting air and the laminated armor glass mounted inside the knife edged forward panels. © *Crown Copyright.*

Top Right. This illustration shows the rear portion of the early style sliding hood, and the rear pressure bulkhead and hood jettison mechanism. © *Crown Copyright.*

Bottom Left. This illustration shows the early-style sliding hood and its multi-piece perspex canopy with vertical and centerline frames. The details of the upper fuselage, cockpit sill and rear pressure bulkhead are also clearly seen. © *Crown Copyright.*

Cockpits - Observer's Cockpit

Left. This image shows the early style observer's hatch in the open position on this FAW 2 . The small window identifies it as such. Note the fabric or leather liner covering the hatch's internal framework, and the window, painted over in black. *Roy Sutherland photo.*

Above. Looking down inside the late style hatch. It has a large window, which in this case also appears to be painted over with black paint. The hinge mechanism is clearly shown. The bracket and red bar holding open the hatch are added by the museum to facilitate access. *Mark Robson photo.*

Left. Looking down on the observer's ejection seat. No parachute, personal survival pack or pads are fitted to this seat. The drogue chute is installed but the cables are disconnected. The bottom of the non-standard door brace can be seen. You can see the equipment rack to the right of the seat. *Mark Robson photo.*

Right. With the seat removed, the starboard wall of "the coalhole" is easily viewed. Called the coal hole because once inside, the observer had little ambient light to work with. Visible in this view is the back of the late style side window, which was hinged at the front and could be opened in a couple of degrees. A bank of instruments include an air pressure indicator and hydraulic pressure gauges. *Roy Sutherland photo.*

Observer's Cockpit continued

With the ejection seat removed, you get a clear view of the rear pressure bulkhead and the back end of the equipment rack. The rear bulkhead features an impressive bank of terminal blocks, capacitors, and a forest of wires and cables. Note the large hot air duct that runs along the bottom of the port canopy sill. *Mark Robson photo.*

The port sidewall of the observer's cockpit. The panel on the top left is the control unit for the observer's oxygen regulator. Top right is a bank of switches for the fuel management system. Below that is an open bay where the in-flight refuelling panel has been removed. You can see the pilot's seat in the bottom right corner. Note the duct mentioned in the previous photo. *Roy Sutherland photo.*

Left. Close-up look at the Mk 17F oxygen regulator and the fuel management panel on the port sidewall. *Mark Robson photo.*

Right. Looking forward at the front cockpit sill from below. Note the large manual locking handle and the two pneumatic hatch locks. *Mark Robson photo.*

Left. Looking down at the observer's office. Tightly packed with equipment, the radar operator had little room to move, and ejection at a moment's notice was not always possible. Note the cover panel on the lower port sidewall is not properly installed. *Roy Sutherland photo.*

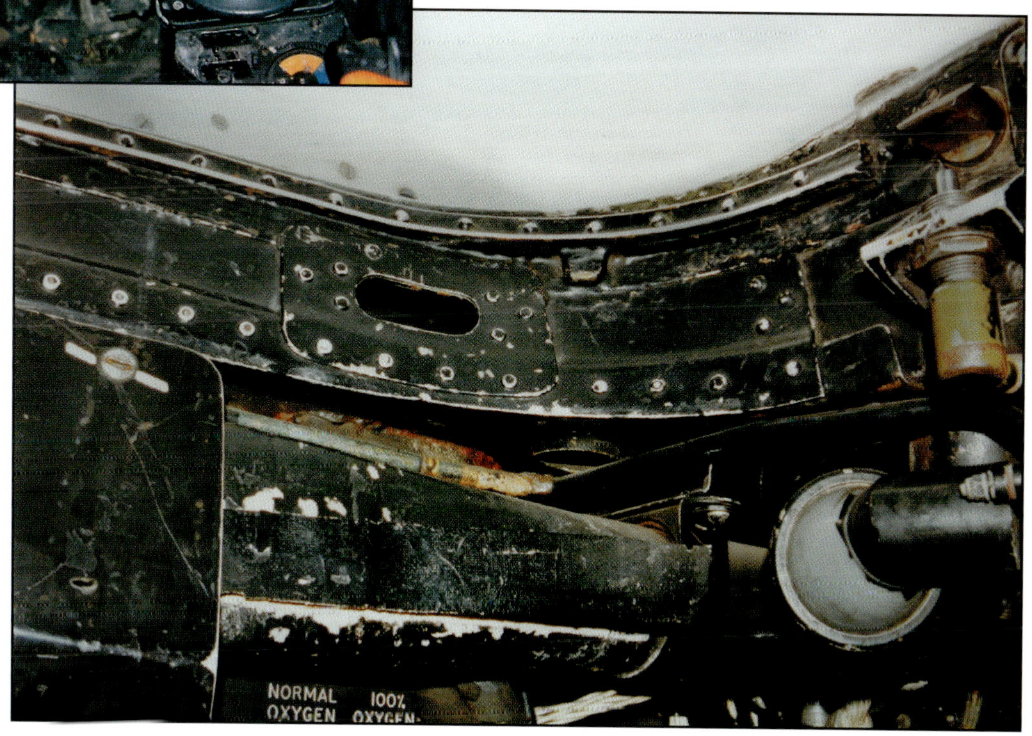

Right. The front left corner of the cockpit sill. Note the recessed handhold, and the heating outlet branching off from the main duct. *Mark Robson photo.*

Right. Central in the cockpit is the display unit (ARI 5950) for the G.E.C. Mk. 18R Airborne Intercept Radar. The two display screens show range and azimuth. Note the rubber visor to shield the screens from outside light, and the pilot's starboard console on the left side of the image. *Roy Sutherland photo.*

Left. The joystick on the starboard side of the observer's cockpit. This can be used to direct the radar scanner under the radome to search a specific quadrant, overriding the set scanning pattern. *Mark Robson photo.*

Right. The monitor meter unit for the ARI 5950 system is at the bottom of the vertical instrument panel. In front of it is the scanner park switch. Above it is a direction indicator. *Mark Robson photo.*

Below. The bottom of the display unit fills the top of the image, and has a number of dials for adjusting the brightness and contrast of the displays. Below this is the control unit for the ARI 5950 system. The white shapes at the bottom corners are the photographers knees, showing just how cramped the observer's office is. Mark Robson photo.

Left. A small instrument cluster above the radar display unit . From left to right are the altimeter, fuel tank display and airspeed indicator. The instrument above them has been deleted. *Mark Robson photo.*

Left. Close up view of the radar display CRTs. On the left is the range/azimuth screen. The right displays azimuth/elevation. The displays are not very bright, necessitating the blocking out of outside illumination. *Mark Robson photo.*

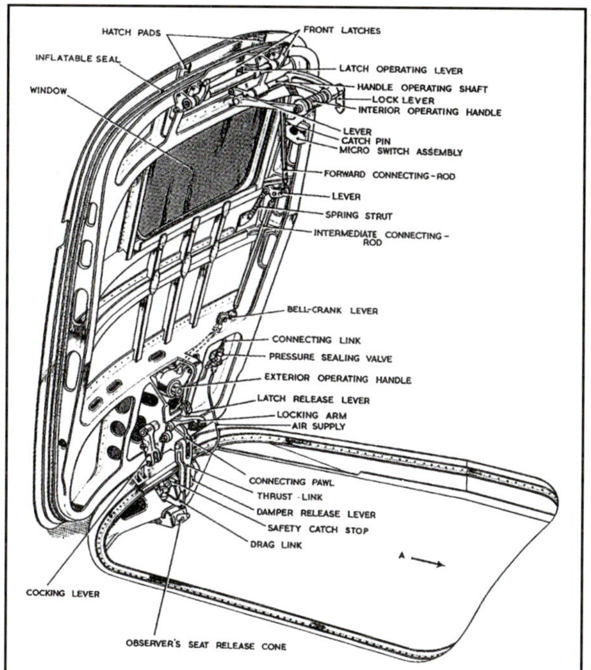

Fig. 11. Observer's hatch
◀Safety claw deleted—Mod. 1550▶

RESTRICTED

Top. A removable window blind is seen partially installed on the observer's window. This view shows the area between the upper instrument panel and the vertical instrument panel. It appears that some loose ejection seat bits have been stuffed in there. *Mark Robson photo.*

Top Right. This illustration shows the structure of the early style observer's hatch. Note that in service, the hatch stays open with a catch mechanism. The inflatable seal is necessary to pressurize the cockpits. A cover appears to be installed over the window. © *Crown Copyright.*

Right. A view seldom seen of the bulkhead that forms the front of the crew cabin. Note the step down on the right for the observers cockpit. © *Crown Copyright.*

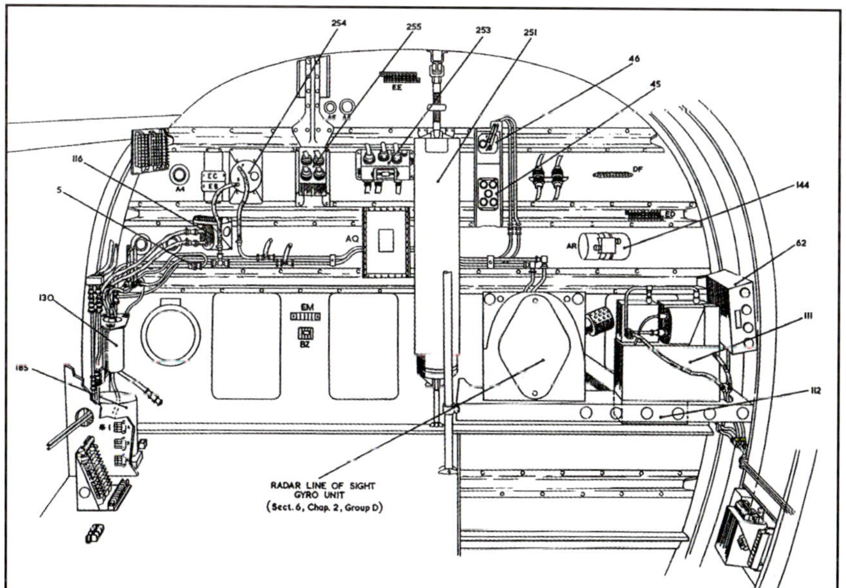

Fig. 11 Rear face, front pressure bulkhead

RESTRICTED

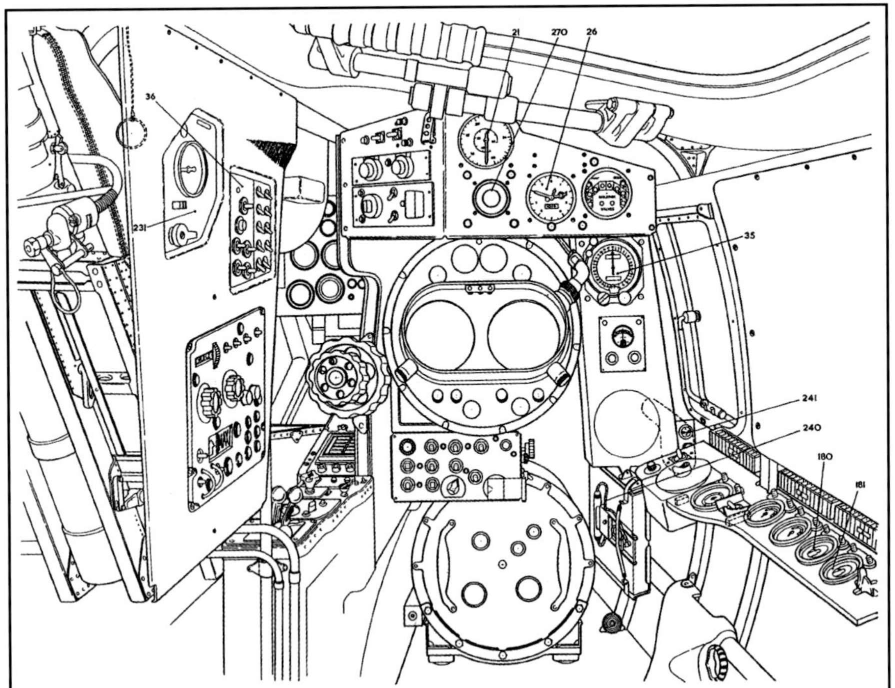

Fig. 9 Instruments and controls in observer's cockpit

RESTRICTED

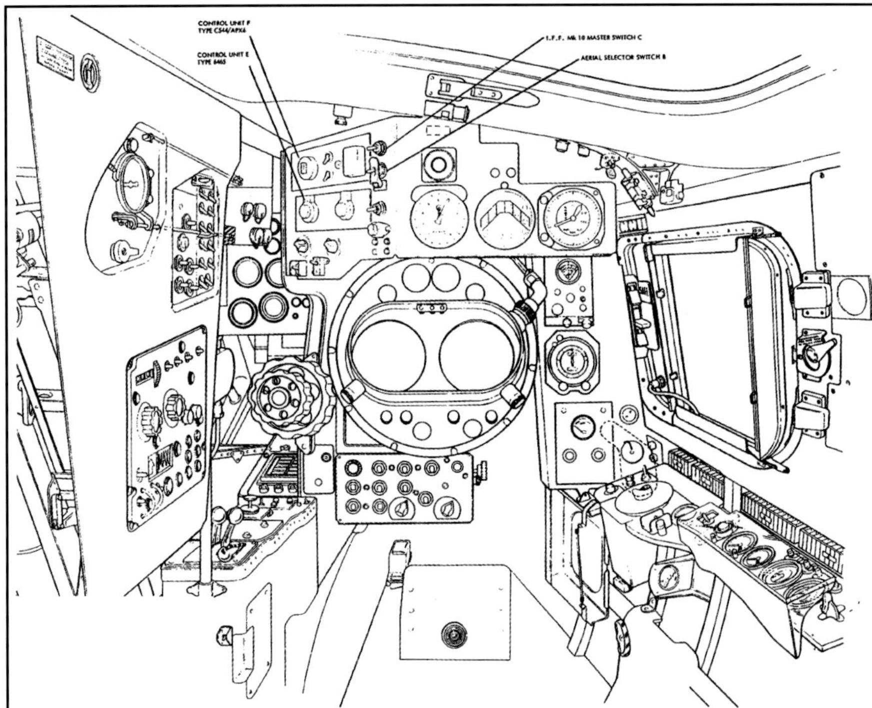

SEA VIXEN FAW Mk. 2 AIRCRAFT

Fig. 1. Controls in observer's cockpit

RESTRICTED

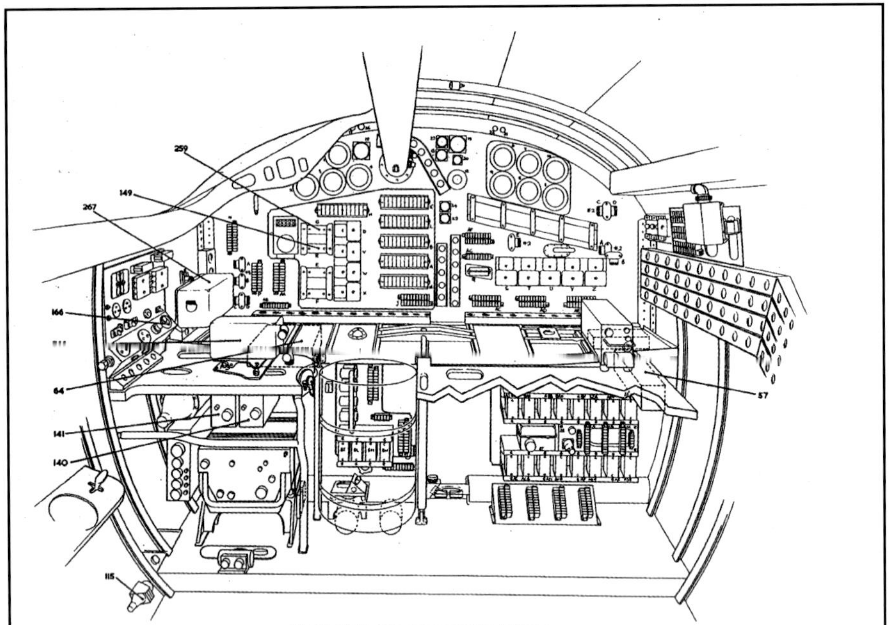

Fig. 12 Front face, rear pressure bulkhead

RESTRICTED

Top Left. A general view of the observers cockpit of the FAW 1, showing the placement of all the pertinent panels and equipment. On the port side wall, the missing bottom panel is shown installed, it appears to be an in-flight refueling control panel. © *Crown Copyright.*

Top Right. A general view of the observers cockpit of the FAW 2, showing the placement of all the pertinent panels and equipment. On the port side wall, the missing bottom panel is shown installed. It appears to be an in-flight refueling control panel. Note the minor differences between the FAW 1 and 2 cockpits. © *Crown Copyright.*

Left. The rear bulkhead, seen earlier in a photo, is shown in its full glory. Most of this will never be seen with the seat installed, but it's interesting to see just how much electrical equipment they managed to pack into a fighter aircraft in the 1960s. © *Crown Copyright.*

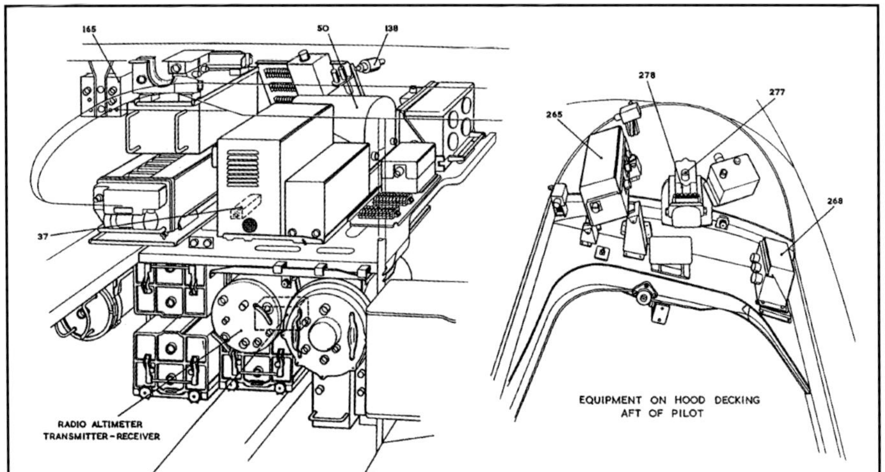

Fig 20 Equipment aft of pilot's seat
RESTRICTED

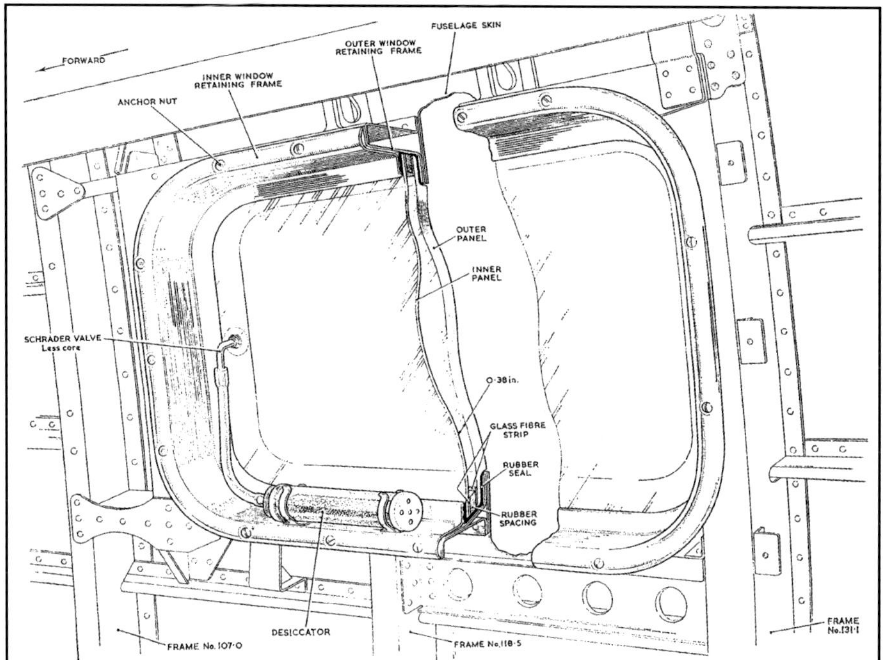

Fig. 14. Observer's window (pre-mod. 1085)
RESTRICTED

Top Left. The equipment rack (consisting of IFF, TACAN, radio and radar gear) that is visible on the left side of the ejection seat is detailed here. Also shown is the rear deck behind the pilot's ejection seat on the FAW 2. This area was less cluttered on the FAW 1. © *Crown Copyright.*

Top Right. The early starboard side fixed window for the observer's cockpit. A fabric cover could be fitted to darken the cockpit, allowing the dim radar screens to be more easily read. © *Crown Copyright.*

Bottom Right. The late style observers side window could be opened inwards a few degrees for ventilation. Doubtful it would ever be opened in flight. It might get a tad breezy in there! © *Crown Copyright.*

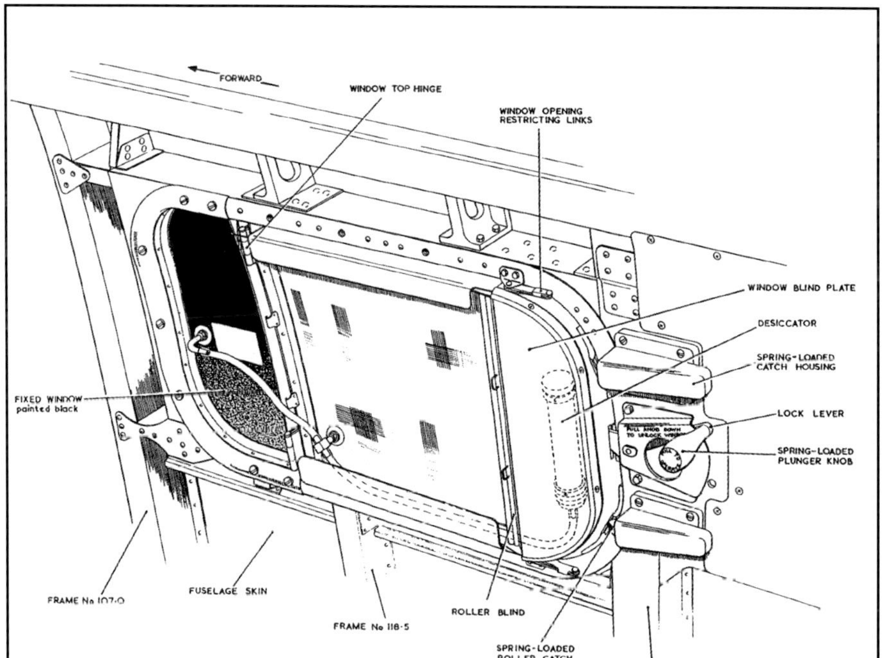

Fig. 15 Observer's window (post mod. 1085 and 1151)
RESTRICTED

Intake and Exhaust

Left. Looking down the port engine intake throat. The turbine face of the Rolls Royce Avon Mk. 208 engine is clearly visible. Each Avon produced 11,000 lbs. of thrust. Note the six centralizer arms ahead of the front row of blades. The intake throat appears to be painted silver. *Roy Sutherland photo.*

Right. Looking down the starboard intake throat to the front of the other Avon. The bullet fairing over the starter is connected to the outer wall by a duct at the 4 o'clock position. This duct carries the pneumatic feed and return lines for the starter. *Roy Sutherland photo.*

Left. Note the contours of the intake in this view and the way the paint demarcation wraps around under the wing. The two guide vanes direct the airflow entering the intake. *Mark Robson photo.*

Right. The boundary layer bleed slot is clearly visible in this shot. Its purpose is to redirect the slow moving turbulent surface air and keep it from entering the engine intake. This slot vents out through openings on the upper surface. *Mark Robson photo.*

Top. This composite image shows the starboard intake cover partially installed. The handle turns the wooden bar on the back of the cover that locks behind the intake vanes to hold it securely in position. *Mark Robson photo.*

Left. The boundary layer bleed slot is clearly visible in this shot. Its purpose is to redirect the slow moving turbulent surface air and keep it from entering the engine intake. This slot vents out through openings on the upper surface. *Mark Robson photo.*

Left. General view of the back end of the fuselage. The two engines were placed so close to the aircraft centerline that single engine performance did not yield asymmetric handling problems. *Roy Sutherland photo.*

Right. The business end of the Rolls Royce Avon Mk 208 axial flow turbojet engine. Avons also powered the Canberra, Swift, Hunter and, in a later variant, the Lightning. *Mark Robson photo.*

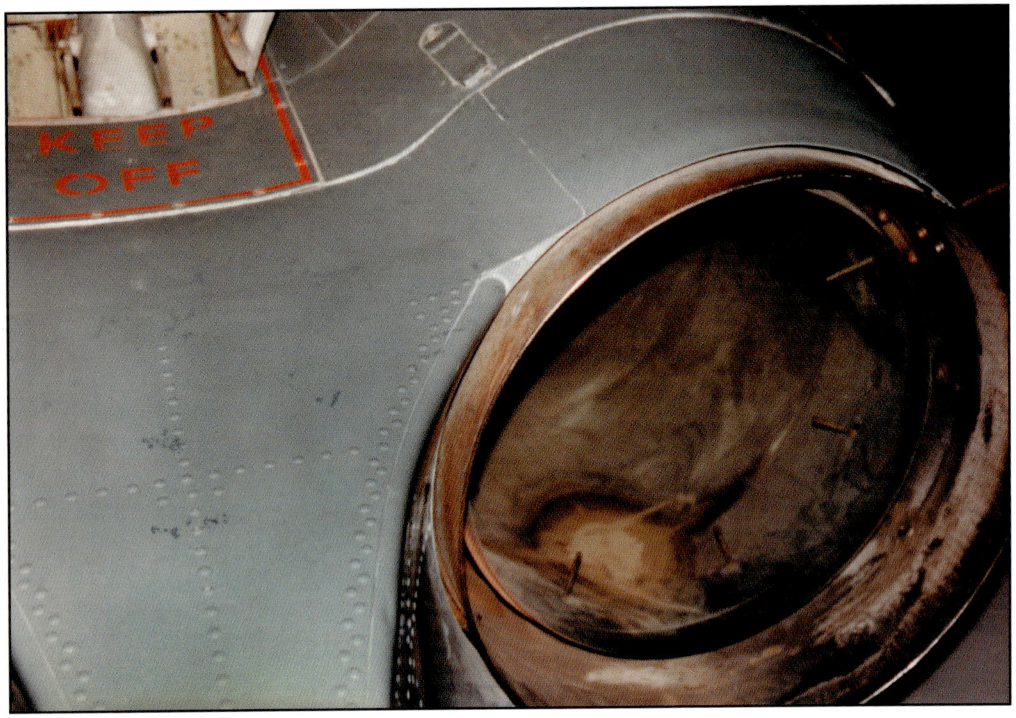

Left. Looking forward and down on the starboard engine exhaust pipe. Note the ring of jet pipe temperature sensors. *Mark Robson photo.*

Right. A Proctologist's view of the Sea Vixen's starboard jetpipe! The Sea Vixen did not have reheat, but was capable of breaking the sound barrier in a shallow dive. *Chris Hughes photo.*

Intake and Exhaust continued

Landing Gear - Nose Gear

Left. A nice 3/4 port view of the Nosegear , showing the early style nosegear door. This door has two forward facing taxying lights and a deck approach light above and between them. *Roy Sutherland photo.*

Right. A look up into the nosegear well shows details of the starboard sidewall. Two hoses feed the declutching valve of the steering mechanism for the nose leg. *Roy Sutherland photo.*

View looking forward shows a blue painted sleeve over the shock absorber strut. This is non standard and could be a support brace. Green numbers 1 and 2 show the port and starboard Microcell launchers in the retracted position. *Roy Sutherland photo.*

This image clearly shows the complex 3 piece nose gear doors. The door with the lights is hinged to the fuselage, while the other two are attached to the gearleg itself. *Roy Sutherland photo.*

Left. Starboard side view of a different Sea Vixen. Note the blade style UHF antenna, and the taxi lights are missing. Note also that this aircraft is not fitted with the retractable microcell rocket launchers, and the resulting differences in the panel lines in this area. *Mark Robson photo.*

Right. Starboard side view of the nose gear. The oleo is uncovered, and the nose gear door has a different arrangement of lights. The stbd. microcell launcher (shown in bright white) is hinged at the rear, and opens a few degrees to allow the launching of the 14 unguided 2 inch rockets. *Mark Robson photo.*

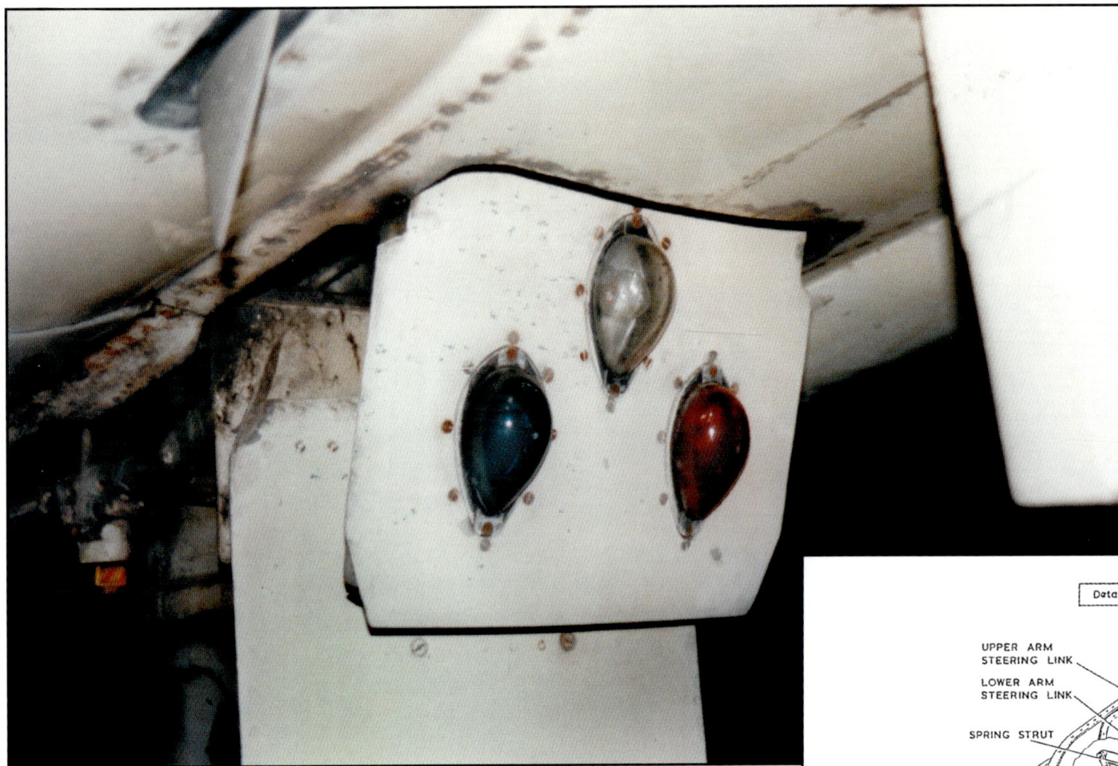

Left. Closeup view of the lights as carried on the later style door. Sea Vixen FAW 1s seem to have mostly been fit with the early style lights, but FAW 2s can be seen fitted with both styles, so check your references. In the upper left corner, the IFF aerial can be seen. *Mark Robson photo.*

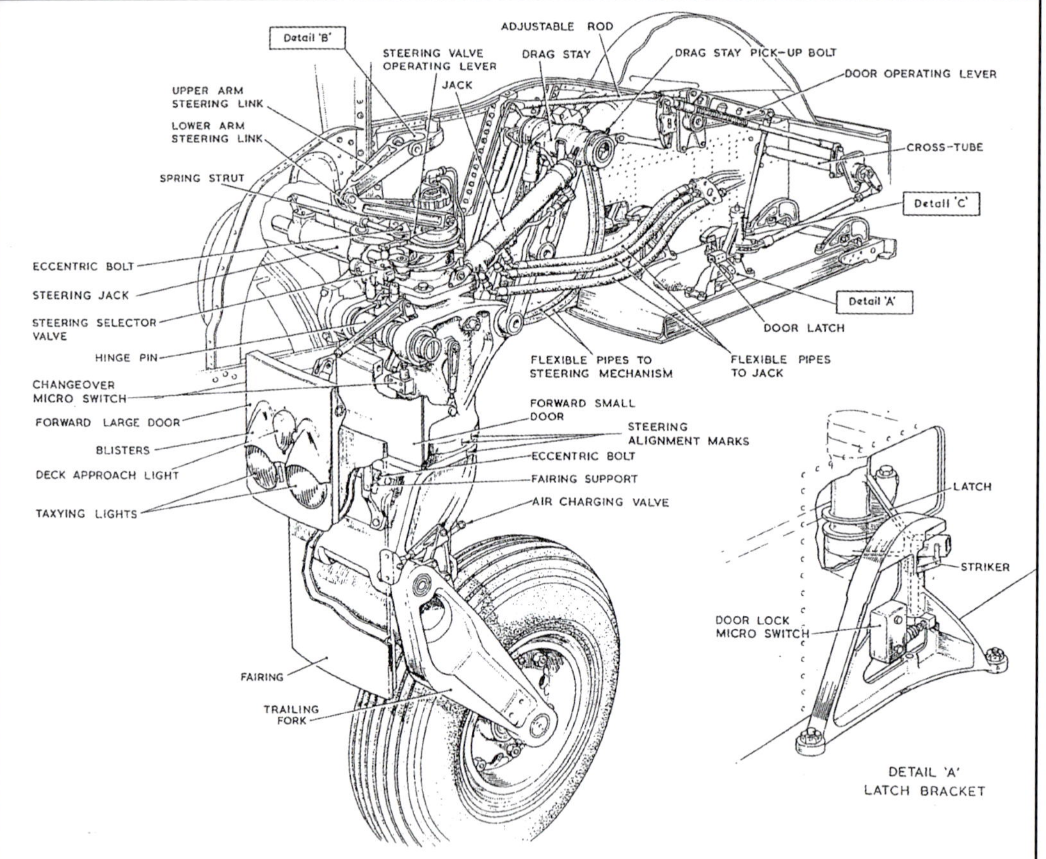

Fig. 4. Nose undercarriage

Right. This excellent illustration shows well the complexity of this nosegear unit. The nose wheel door remains shut when on the ground. Hinged at its rear edge, it opens only during retraction to allow the gear to enter the bay, then closes again afterwards. © *Crown Copyright.*

Landing Gear - Main Landing Gear

Looking aft at the port maingear leg and wheel. The extremely heavy duty nature of the Sea Vixen's undercarriage is evident in this view. Landing almost 24 tons of jet fighter on a short, pitching carrier deck demands a lot of them. *Roy Sutherland photo.*

This image of the port maingear clearly shows the hefty side stay. This brace is also connected to the retraction jack. The attachment point for the door closing strut and the serpentine brake cable runs are also easily seen in this view. *Roy Sutherland photo.*

The starboard main gear , as seen from behind, shows the seldom seen back of the maingear leg assembly. Note the deeply dished brake side of the wheel, and the interesting contours of the leg door. *Roy Sutherland photo.*

Left. A great view of the substantial torque link and the triple brake lines. The three lines are for pressure feed and return on the main brakes, and the pressure line for the parking brake. *Chris Hughes photo.*

Right. Looking outboard at the port mainwheel of the Sea Vixen. This is the later so-called Sea Vixen wheel. Earlier Sea Vixens were fitted with mainwheels identical to those designed for the Supermarine Scimitar. Dunlop 35" tires fitted. *Roy Sutherland photo.*

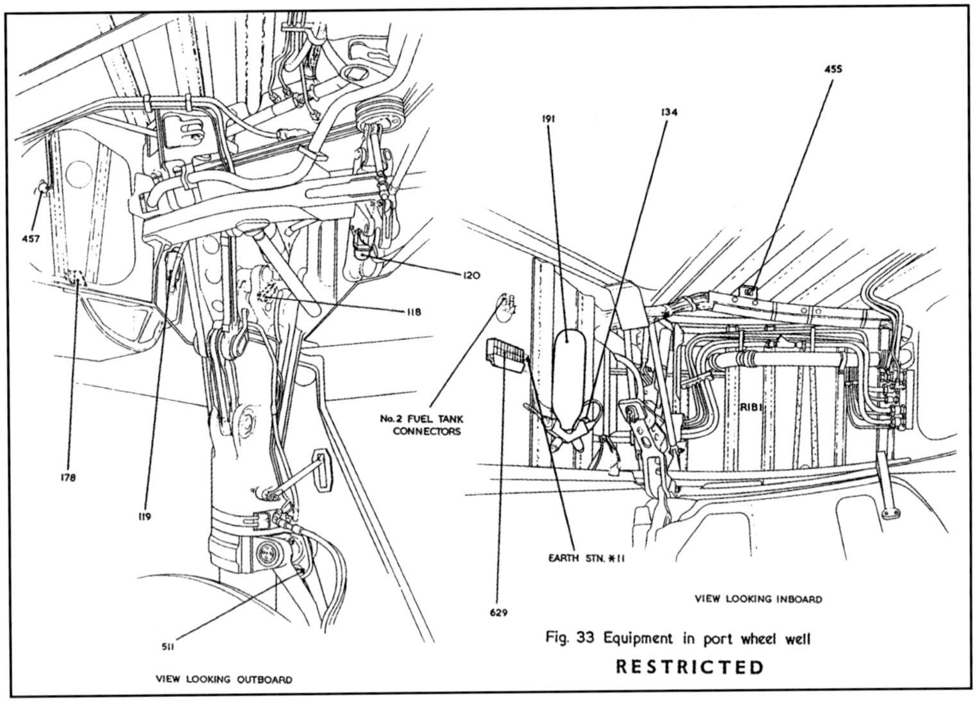

Fig. 33 Equipment in port wheel well

RESTRICTED

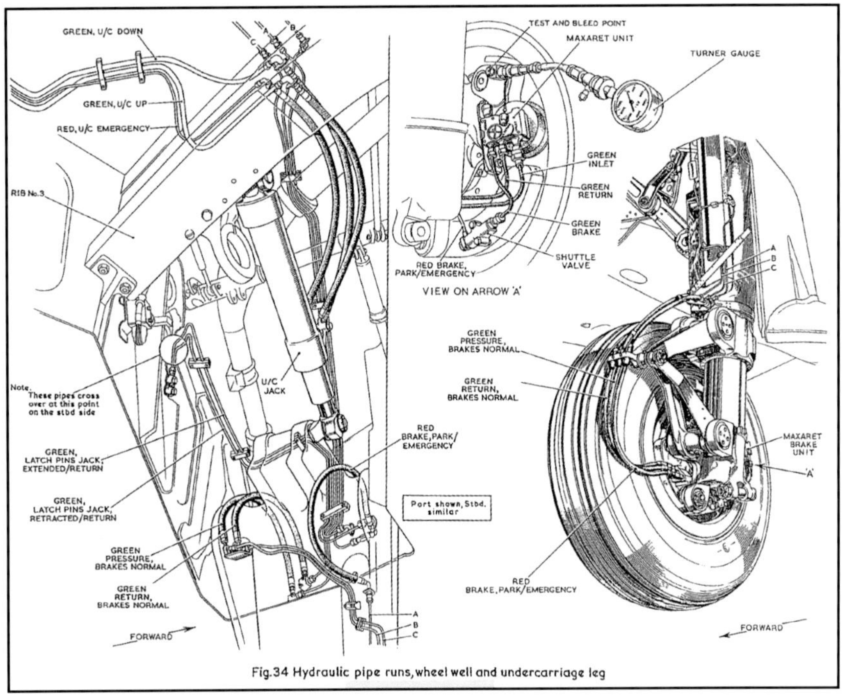

Fig.34 Hydraulic pipe runs, wheel well and undercarriage leg

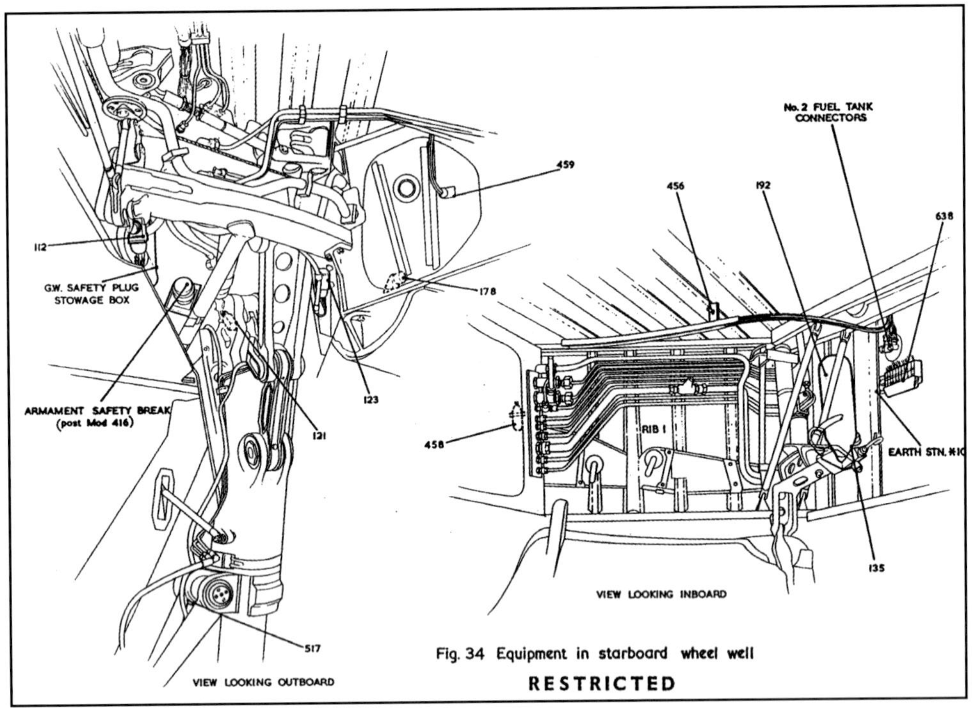

Fig. 34 Equipment in starboard wheel well

RESTRICTED

Top Left. This drawing is useful for making out structures in the sometimes overwhelming detail photos of the upper gearleg and gearbay. Port shown. © *Crown Copyright.*

Top Right. For the modeller who wants to get every detail right, this illustration reveals the complicated brake lines, from the brake unit in the wheel hub up into the gearbay. Maxaret brake units are an early anti-lock system to prevent skidding. © *Crown Copyright.*

Bottom Left. This drawing is useful for making out structures in the sometimes overwhelming detail photos of the upper gearleg and gearbay. Starboard shown. © *Crown Copyright.*

Landing Gear - Main Gear Bays

Left. The front wall of the port gearbay is formed by the wing center spar. Note the linkage between the maingear and the wheel door retraction mechanism. Gearbays were painted gloss white. *Roy Sutherland photo.*

Right. Looking up into the roof of the port main gearbay. The stencil reads “These four access panels must be secured before flight”. Starboard gearbay is similar but with slightly different plumbing. *Roy Sutherland photo.*

Above. A view looking outboard in the port main gearbay. Prominent is the notched wing rib 2. The retraction jack and side stay are mounted to this substantial member. *Roy Sutherland photo.*

Above. Again, the port gearbay, this view showing the rear wall, and two of the four access panels referred to in the lower photo on page 75. *Roy Sutherland photo.*

Facing Top Left. Looking aft and inboard in the port gearbay. The large mainwheel door remains closed on the ground, and is normally open only during retraction and extension. *Roy Sutherland photo.*

Facing Bottom Left. The inboard and forward port gearbay walls and the mass of hydraulic and electrical runs are evident in this view. Note the wheel door retraction jack and the retraction linkage. *Roy Sutherland photo.*

Above. Another view outboard of the port main gearbay. This view shows more of the gearbay roof, showing to better effect the wiring runs that circle the bay. *Roy Sutherland photo.*

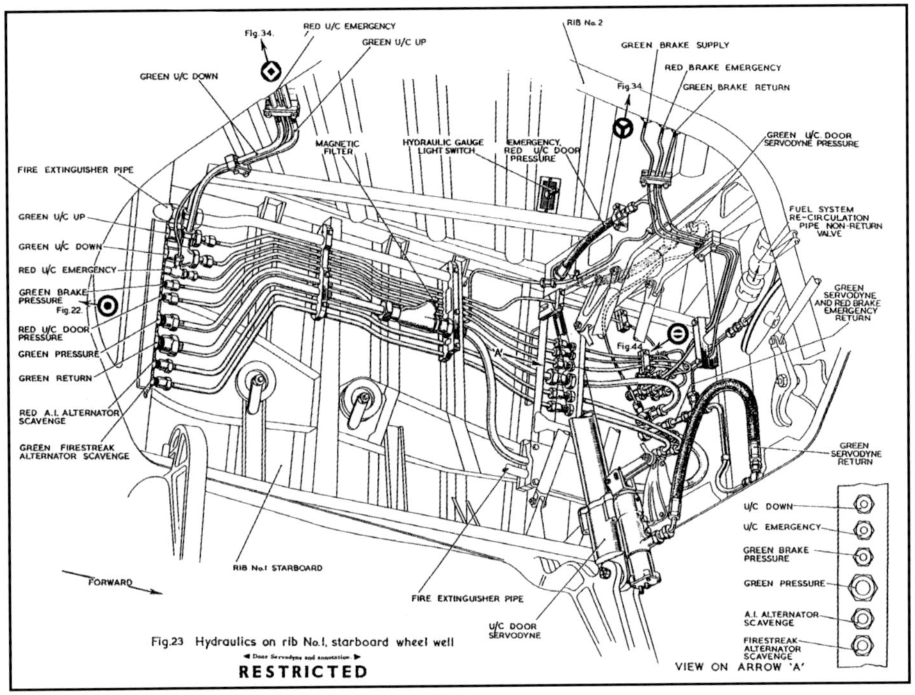

Fig.23 Hydraulics on rib No.1, starboard wheel well

RESTRICTED

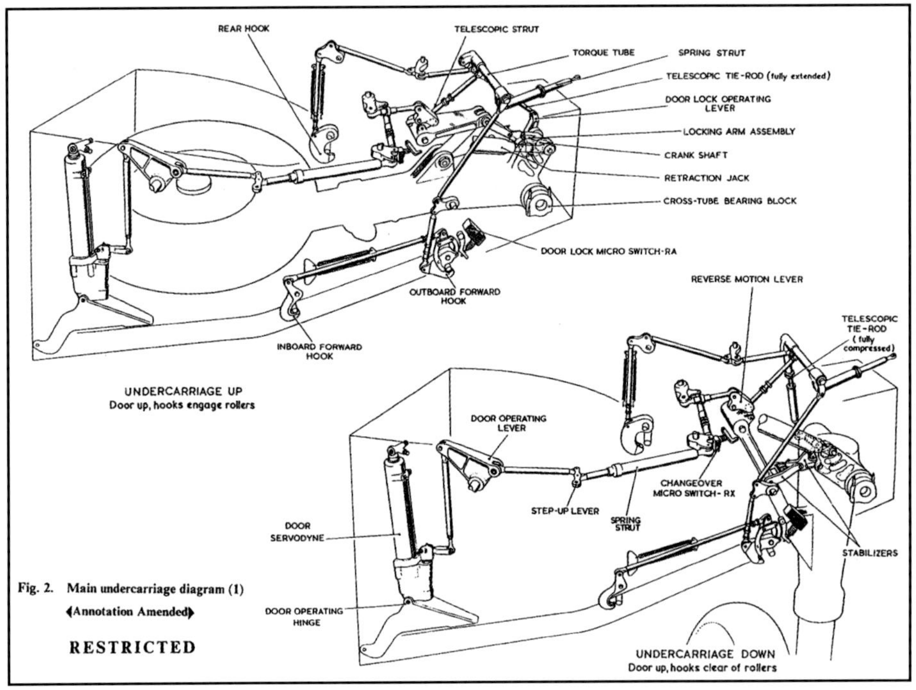

Fig. 2. Main undercarriage diagram (1)

(Annotation Amended)

RESTRICTED

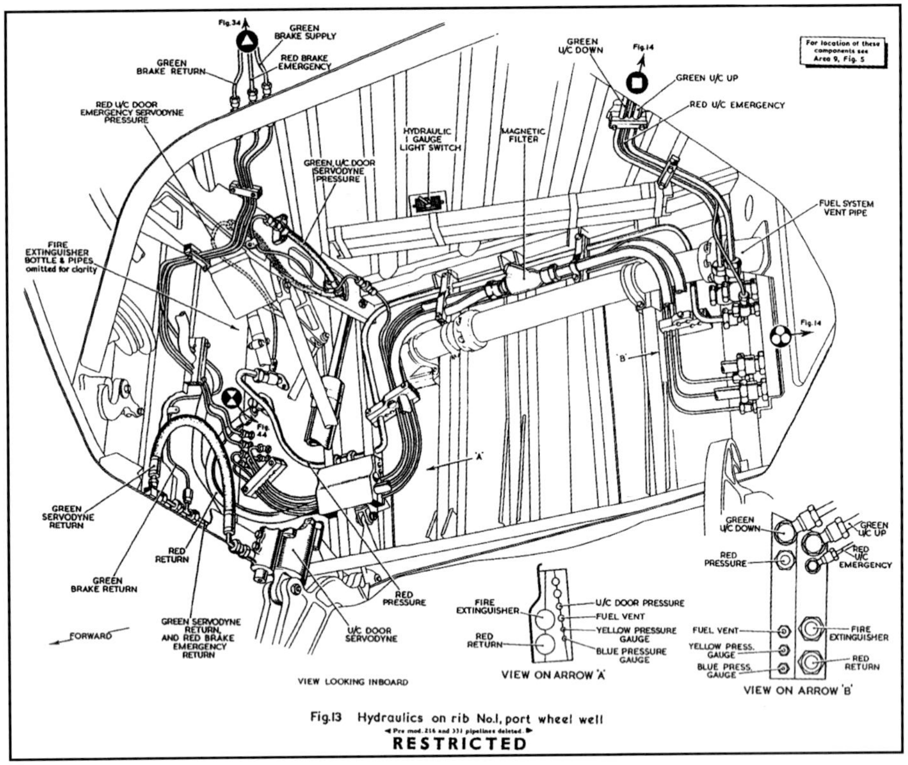

Fig.13 Hydraulics on rib No.1, port wheel well

RESTRICTED

Top Left. This illustration details the mass of plumbing in the Sea Vixen's starboard gearbay. © *Crown Copyright.*

Top Right. A useful diagram showing the intricate system of pistons, linkages, pushrods and bellcranks that comprise the retraction system of the maingear. © *Crown Copyright.*

Bottom Left. This illustration details the mass of plumbing in the Sea Vixen's port gearbay. © *Crown Copyright.*

Landing Gear - Arrestor Gear

Above. Looking up and forward into the arrestor hook well. Notice the mounting straps for two compressed gas cylinders, presumably for emergency operation of the arrestor hook . Check out the oddly shaped drain in the lower left corner, designed to keep fuel from flowing back into the jet efflux. *Roy Sutherland photo.*

Right. Looking forward at the upper portion on the arrestor hook and the hook fairing. The retraction jack is missing, but the hook centering mechanism is installed. The fabric strap is non-standard. *Roy Sutherland photo.*

Arrestor Gear

This shot shows the aft end of the hook well. Unlike the rest of the Sea Vixen airframe, this area is simple and uncluttered. *Roy Sutherland photo.*

This image shows the entire hook and well as viewed from directly behind. The position of the aft end of the hook well in relation to the tail fairing between the exhausts is shown. *Roy Sutherland photo.*

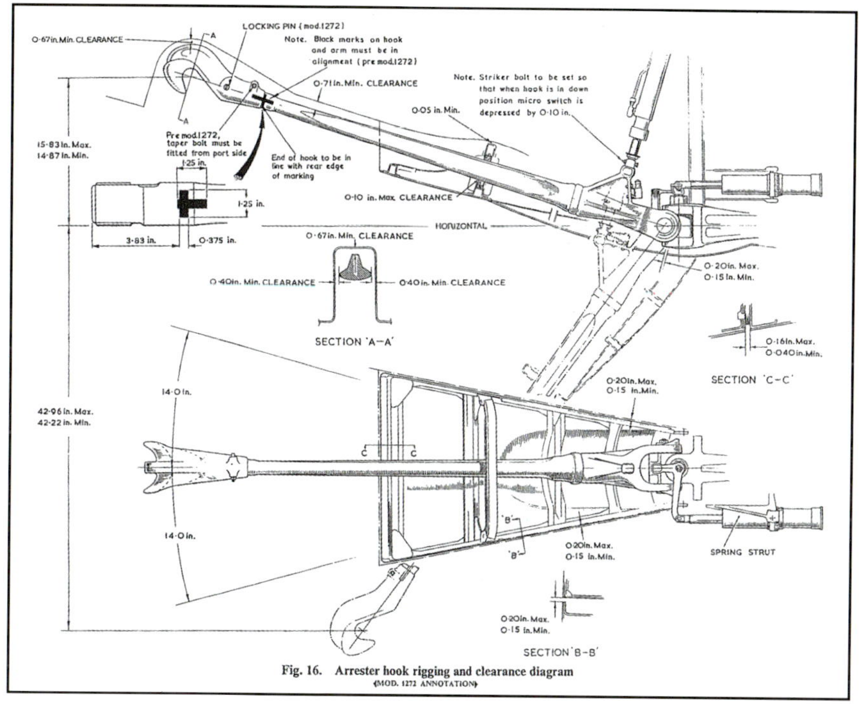

Fig. 16. Arrester hook rigging and clearance diagram
(MOD. 1272 ANNOTATION)

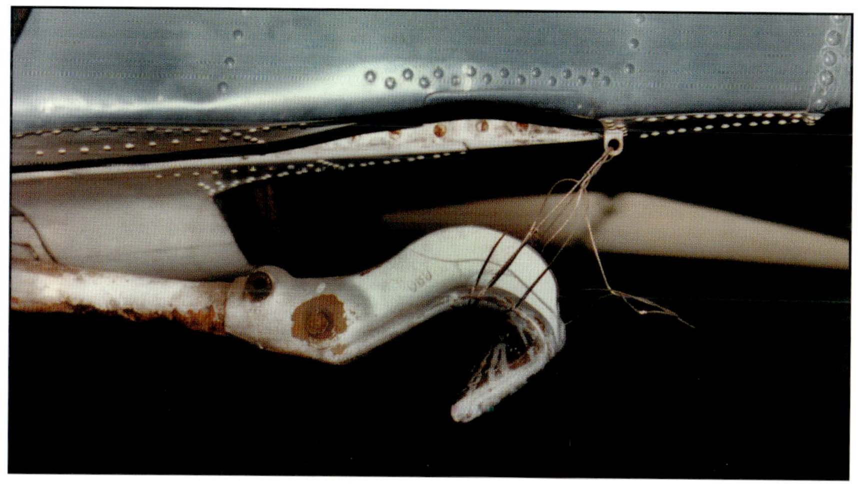

Above. Another view of the tailhook and the pen nib fairing between the exhausts. *Mark Robson photo.*

Top Right. This drawing shows the retraction jack attached, and the spring jack that automatically centers the hook when it is detached from the arrestor cable. © *Crown Copyright.*

Above. Closeup port side view of the business end of the arrestor system. The hook is a forging. The hook is obviously jury rigged with wire to keep it from resting on the ground. *Mark Robson photo.*

Section continued

Wings - Outer Wings

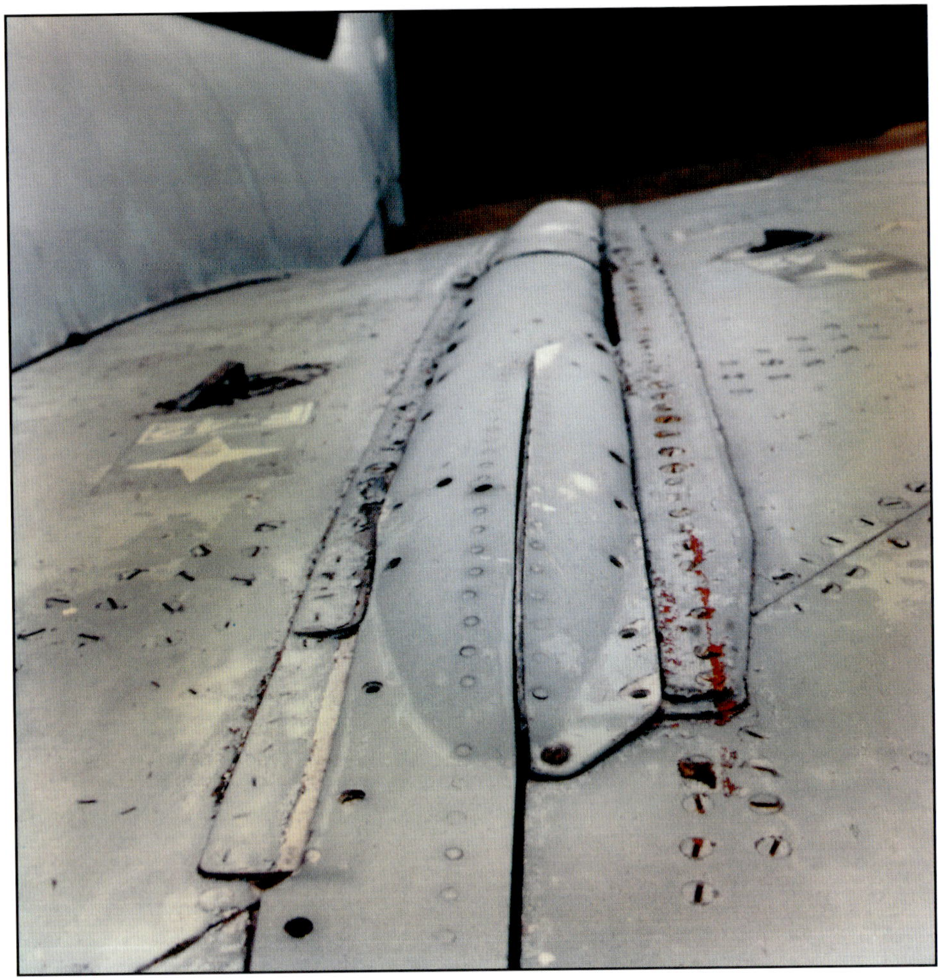

A close up view of the multi-part port wingfold fairing. This view looking aft shows a large number of screws used in this area of the wing. Note the inner and outer wing fuel tank filler points. *Mark Robson photo.*

The rear view of the same port wing fairing. Note the spring loaded hinged sections that allow the hinges clearance when the wings are folded. Also visible on the left side is another filler cap for the outer wing tanks. *Mark Robson photo.*

Right. Looking inboard and forward across the top of the starboard wing. The long flattened tube is a fuel dump. The rubber tip on this aircraft is sun damaged. It should be black. The small access panel is for the flap uplock mechanism. Note the cutback wing that exposes the trailing edge of the flaps. *Mark Robson photo.*

Bottom Right. The outer face of the port wing fence. Wing fences are designed to keep air from moving span-wise at lower speeds. This is a problem on swept wings as it decreases lift and causes pitch up and tip stalls. This can lead to possibly deadly situations during slow flight and landing. *Mark Robson photo.*

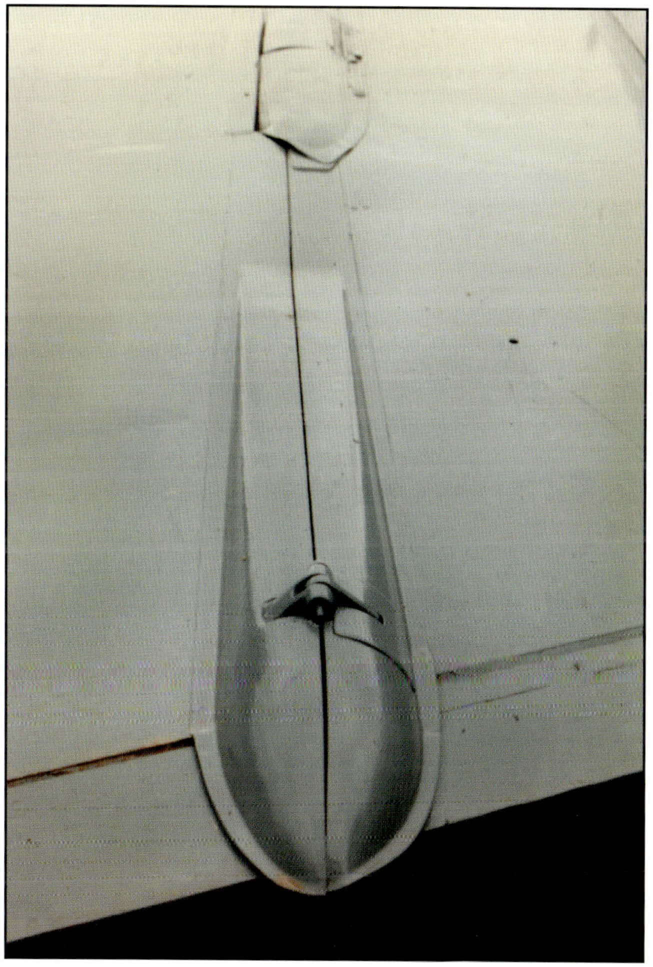

Left. The starboard trailing edge flap guide fairing viewed from the rear. Note the raised hinge, necessary to keep the pivot point in line with the other wingfold hinges. Mark Robson photo.

Outer Wings

Top Left. The starboard wingtip navigation light. Starboard lights are green, while the light in the port wingtip would be red. *Mark Robson photo.*

Top Right. Looking forward at the port wing fence. A honeycomb structure lies beneath the skins of the Sea Vixen's wing fences. L-shaped mounting strips secure the fence to the wing. *Mark Robson photo.*

Bottom Left. A closeup of the pitot tube on the folded starboard wing. The Sea Vixen has a similar pitot tube on the port wing. The pitot tube's function is to measure airspeed. *Mark Robson photo.*

Facing Top Left. A somewhat poor photo, but included for its interesting content. This FAW 2 has had its starboard pintle tank removed, exposing the original Mk 1 boom, and the hardware added to it to mount the extended booms of the FAW 2. Also noteworthy is the jury strut mounting on the outer wing and the missing wing fold fairing panels. *Roy Sutherland photo.*

Above. The inner face of the port wing fence. The fence was installed at the wing leading edge dog-tooth, another device to improve low speed flight characteristics. Swept wings work very well at high speeds. *Mark Robson photo.*

Left. The lower surface of the starboard wing, showing details of the aileron and its hinges as well as the position of the underwing roundel and serial. *Mark Robson photo.*

Left. The mass balance of the huge starboard aileron is shown to good effect here. Inset photo shows an additional light towards the rear of the aileron. Color and purpose unknown. Possibly a position light? Note also the small drain at the tip. *Mark Robson photo.*

Bottom Left. This technical illustration shows the surprisingly large number of electrical components installed in the outer wings. © *Crown Copyright.*

Bottom Right. The port outer wing laid bare. Rib 4A is machined out of a solid billet of aluminum for strength. The wing skins on the Sea Vixen were chemically milled so that they taper in thickness towards the wingtip. In fact, many of the skins were so treated. © *Crown Copyright.*

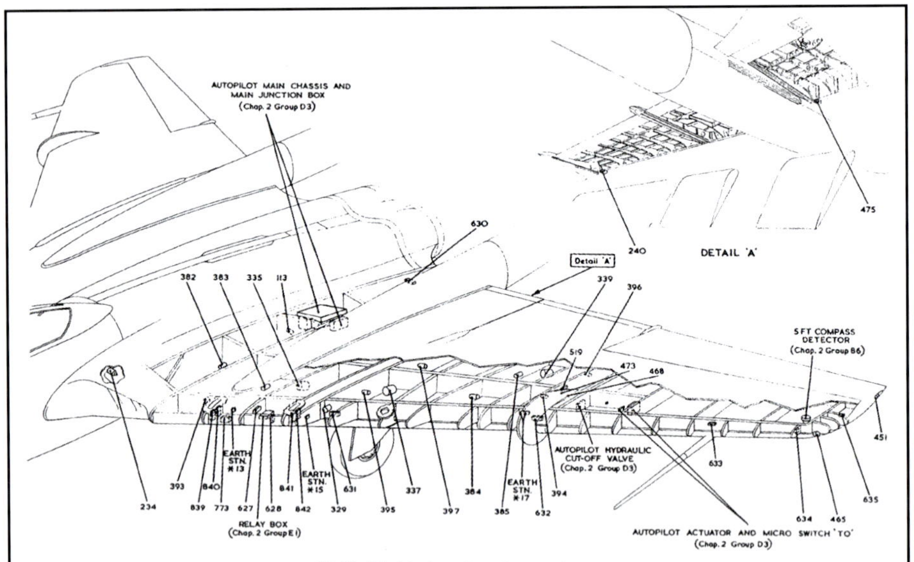

Fig.26 Electrical equipment port wing
RESTRICTED

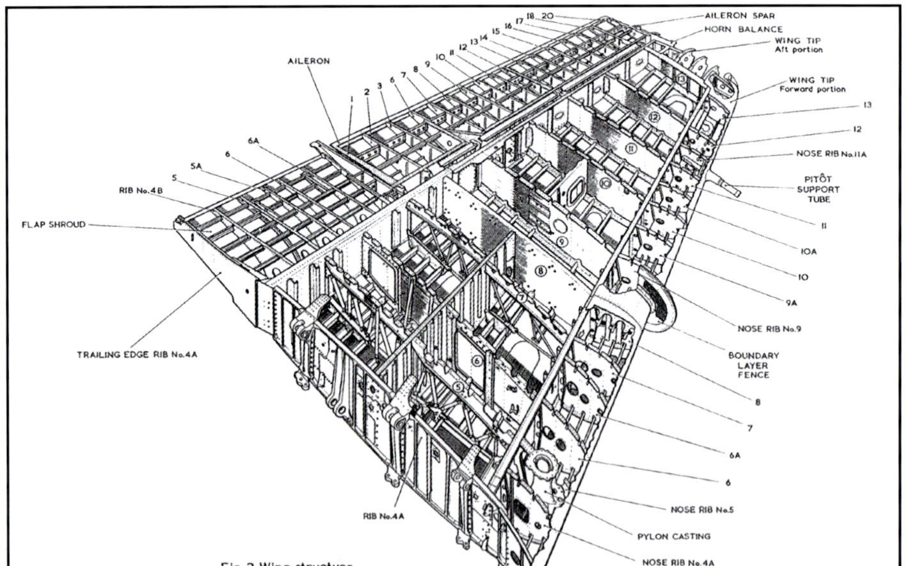

Fig.3 Wing structure
RESTRICTED

Wings - Wingfolds

Above. The front portion of the Sea Vixen's port wingfold. The large white chute is a folding metal guide for the many electrical cables that provide power to the outer wings. Behind that are the single horizontal and two vertical pins that lock the wing in the spread position. Note the spigot and bush (far left) that further stiffen the spread wings. *Roy Sutherland photo.*

Top Right. On the left is the complex system of pushrods and bellcranks that operate the locking mechanism. Two more locking pins are also visible. In the center are the two swiveling fuel tank pipes. One is for feed and the other for venting. Behind that is the piston and mechanism for power folding the wings and a battery of hydraulic hoses. *Roy Sutherland photo.*

Right. A great shot looking back at the entire port wingfold. This view gives a more 3 dimensional idea of the various components, especially the power wingfold mechanism *Roy Sutherland photo.*

Wingfolds

Top Left. A slightly different view of the wingfold center section. Note the jack that operates the locking pins (in the bottom left corner), the center locking pins and directly above them, the forward wing hinge. *Roy Sutherland photo.*

Top Right. Yet another view of the forward portion of the wingfold. I find it very helpful to study a complex assembly from many different views to help understand how it looks and works. Both wing ribs are milled from solid aluminum billets, and are very sturdy . *Mark Robson photo.*

Left. Aft of the wingfold operating mechanism are the flexible hydraulic hoses that control the aileron. These lines then run aft on the outboard wing section. Note the rear wingfold hinge and the 2 rear locking pins beneath it. *Mark Robson photo.*

Top Left. A better view of the 4 hydraulic aileron control lines running chordwise. On the right is the rear wingfold hinge. Aft of them are two sequencing valves. The circular structure in the center of the image is the flap drive connection to the outer wing. *Mark Robson photo.*

Bottom Left. The trailing edge of the starboard wingfold. This is the wingfold flap guide fairing in the folded position. You can just make out the return sprocket for the flap chain drive in both wing sections. Note the protruding outer flap's interlock arm. *Mark Robson photo.*

Above. Viewed from directly behind, the angle of the starboard wing when folded is demonstrated. Note the flap guide hinge in the folded position, the two open wingfold hinge fairing doors, and a great shot of the wing tank fuel dump. *Mark Robson photo.*

Left. The wing in folded position showing the jury strut that holds the wing securely when folded. Jury Struts are especially necessary on windswept Carrier decks to prevent damage. *Roy Sutherland photo.*

Bottom Left. Just how complicated the Sea Vixen's wingfold is can be seen in this detailed diagram. © *Crown Copyright.*

Bottom Right. The inner wing fold stub with the outer folding wing removed. This shows the hinges and the hinge fairing doors clearly. Note the disconnected electrical and hydraulic cables. © *Crown Copyright.*

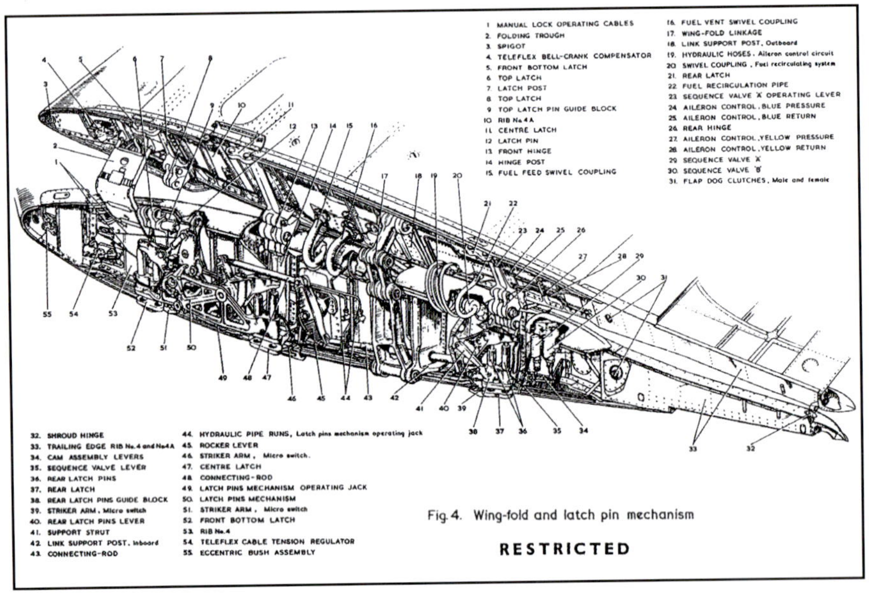

Fig. 4. Wing-fold and latch pin mechanism

RESTRICTED

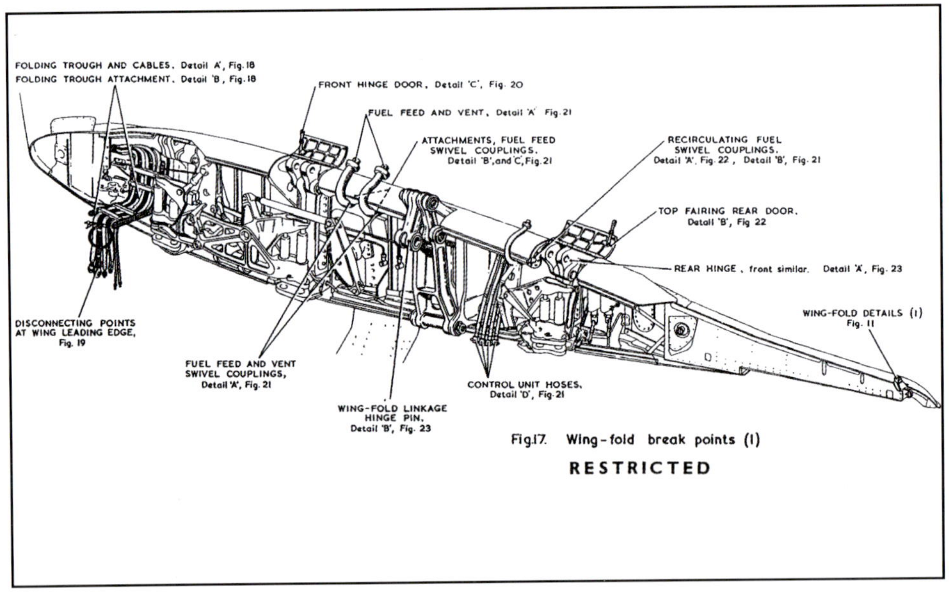

Fig. 17. Wing-fold break points (1)

RESTRICTED

Left. View up into the port outer flap well. The flaps on this Sea Vixen seem to be disconnected from the tracks, as they are fully forward, but in the down position. Fowler flaps deploy as they slide back on tracks. Note non-standard broken tree branch laying on top of flap. *Mark Robson photo.*

Right. Port outer flap again, but this time looking outboard. The bulged flap fairing is visible, as is the chain drive that moves the flaps into the desired positions. Note the flap roller on the upper surface of the flap that catches the uplock mechanism when retracted. *Mark Robson photo.*

Top Left. The inboard portion of the port outer flap is shown in this image. Note the flap interlock mechanism protruding from the side of the flap near the trailing edge. The spider web, as with the broken branch, was not part of the original design! *Mark Robson photo.*

Top Right. The center portion of the port outer flap well. Details are mostly identical to the starboard outer flap and well. The inner flaps were not deployed on any Sea Vixen photographed, but the general look is very similar. *Mark Robson photo.*

Bottom Left. The same area as in the above image, but looking inboard. Note the plumbing running along the forward wall of the flap well, and the flap drive chain on the sidewall. *Mark Robson photo.*

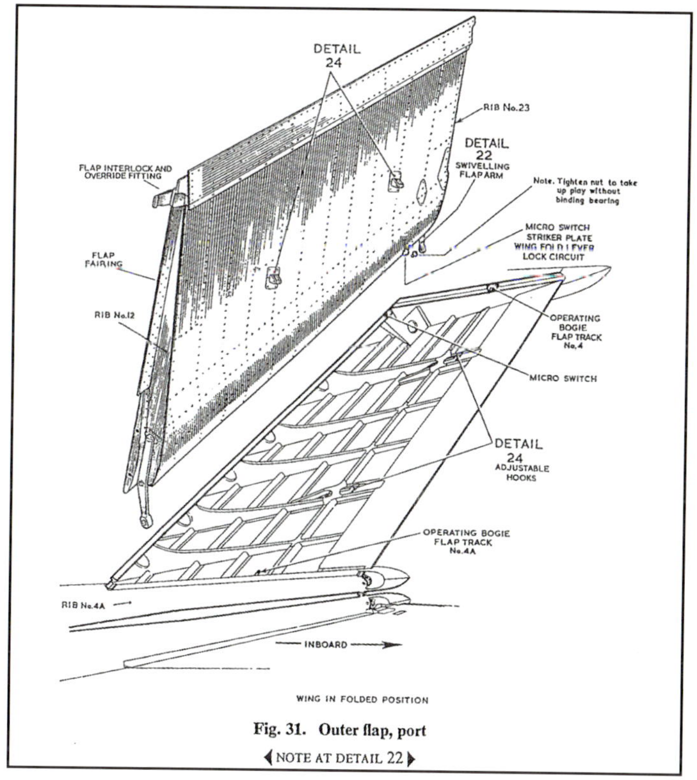

Fig. 31. Outer flap, port

NOTE AT DETAIL 22

Above. This in-flight shot shows the port flaps partially deployed. When fully deployed, the leading edge of the flap is pretty much lined up with the trailing edge of the wing. © *Peter Steehouwer photo.*

Top Right. This drawing can be used with the accompanying photos to get an excellent idea of what the Sea Vixen's outer flaps looked like. The upper wing skin is cut back, exposing a few inches of the upper trailing edges of the middle and outer flaps when retracted. © *Crown Copyright.*

Right. Technical illustration of the complete flap system. The U-plates fit flush around the bottom of the booms when retracted. Note the micro switch that prevents the wings from being folded when the flaps are deployed. The elevator is connected to the flaps via a cable to automatically compensate for trim changes. © *Crown Copyright.*

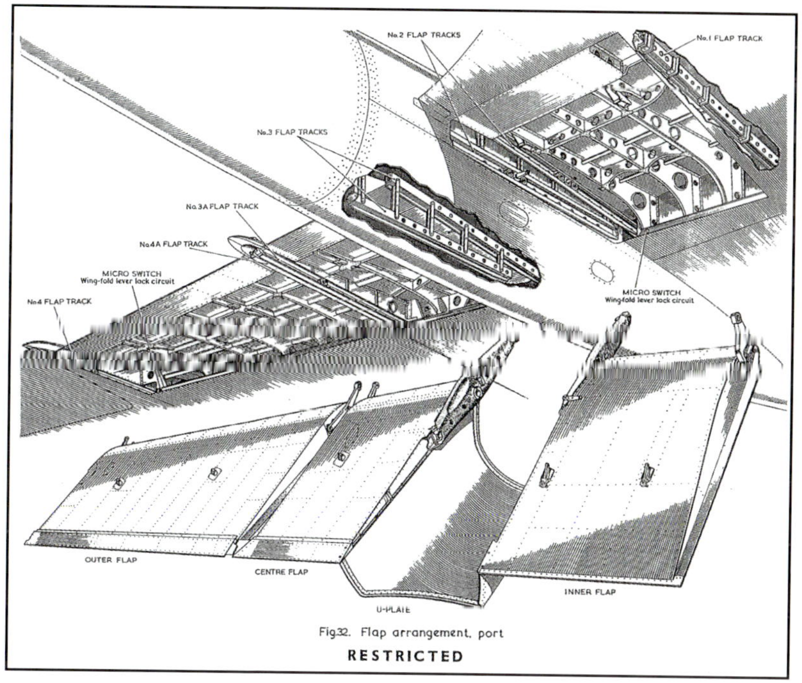

Fig32. Flap arrangement, port

RESTRICTED

Wings - Weapons and Tanks

Wings - Weapons and Tanks

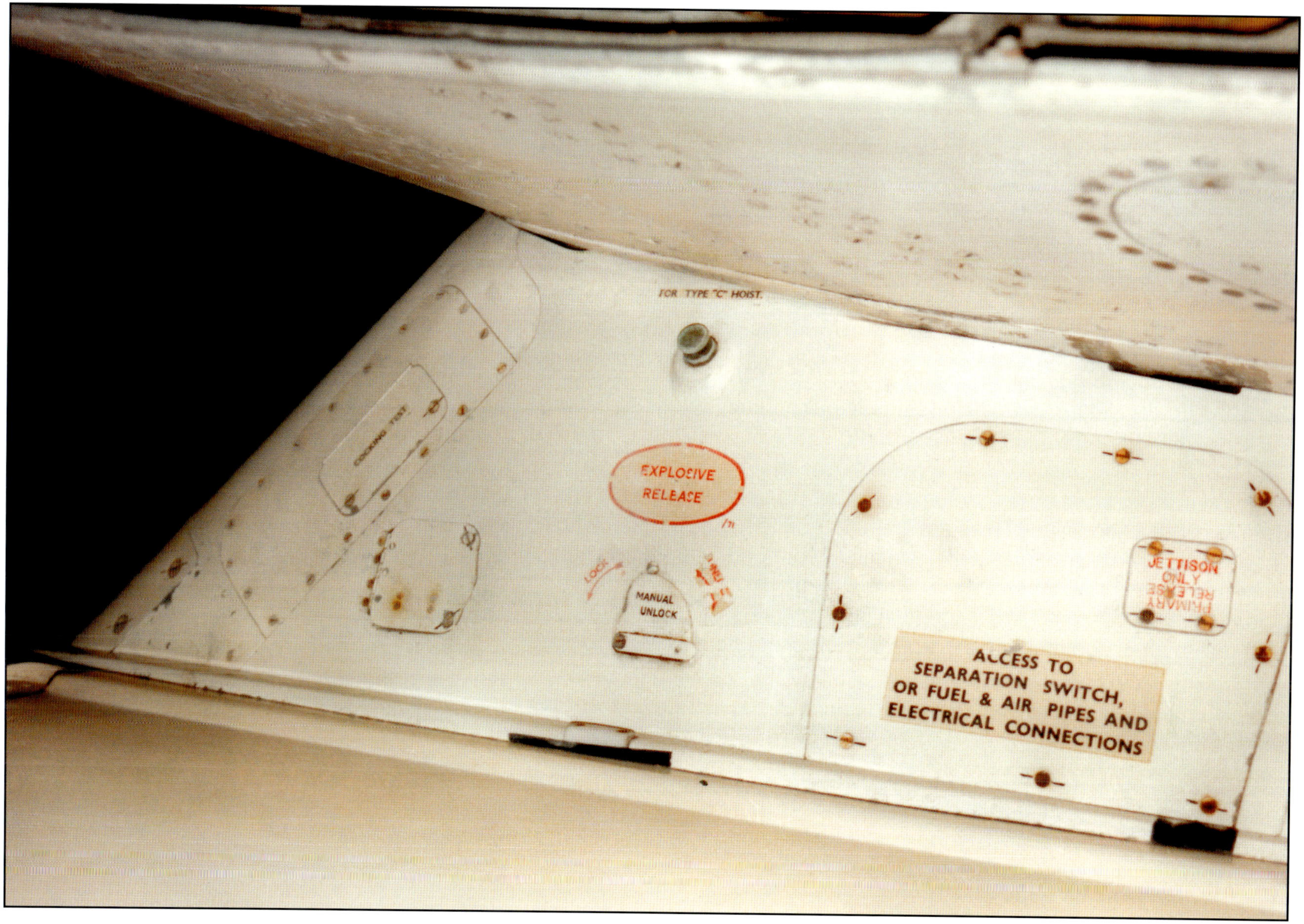

Facing. Looking outboard at the starboard droptank and pylon on the outer wing. The image is inverted for clarity. This droptank is one of a pair mounted on a Sea Vixen with its wings folded, so the tanks were upside down 10 feet off the ground, and not easy to photograph. *Mark Robson photo.*

Above. A closeup look at the inner face of the starboard droptank pylon. Note the various access panels and that all stencils are decals. *Mark Robson photo.*

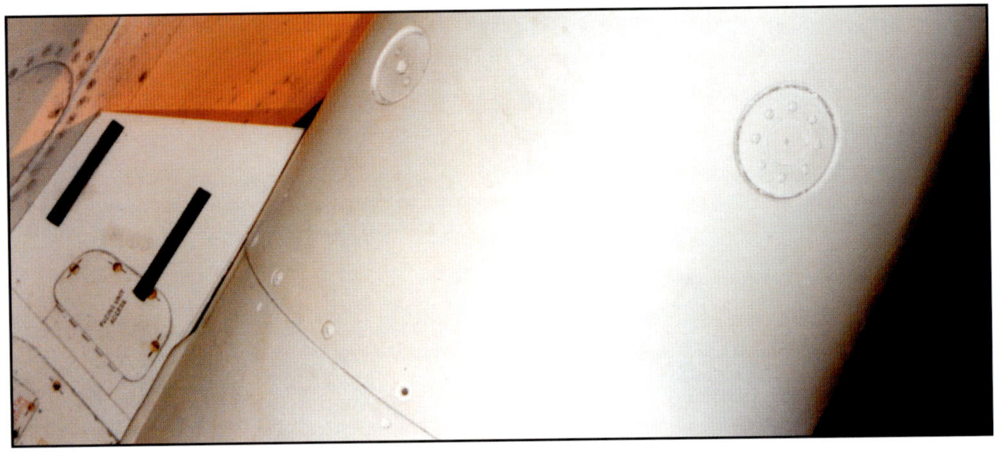

Above. An odd angle shows the rear portion of the droptank. Each droptank holds 150 imperial gallons, feeding the thirsty Avons to extend the operational range. *Mark Robson photo.*

Below. Under the wire! This last minute addition shows the whole droptank to good advantage. This beautifully restored Sea Vixen shows a few compromises to crowd pleasing. The droptanks are mounted to the inner light stores pylon. This is not possible on an operational aircraft, as there are no fuel connections there. *Peter Woolley photo.*

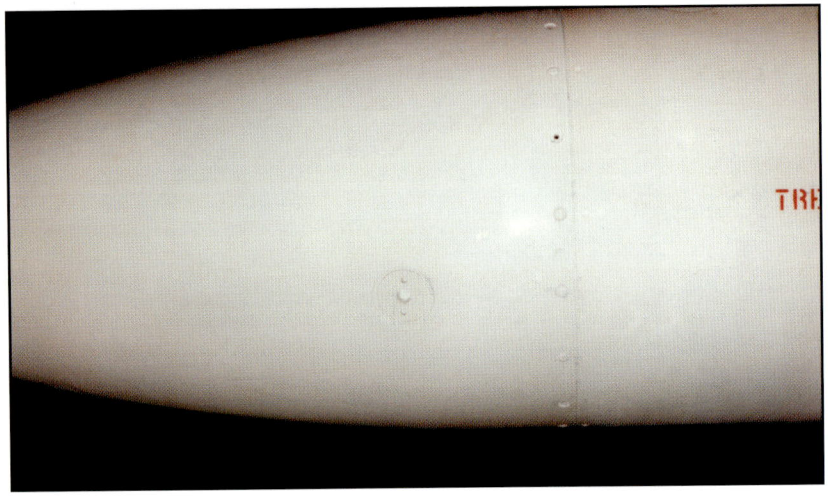

Above. The front nosecap of the starboard droptank. Mark and I shot four Sea Vixens between us, yet I have come up short on photos of the droptanks. Only one was fitted with them, and somehow this is all the shots we managed to take. *Mark Robson photo.*

Below. This shot shows the rear end of the raised area surrounding the pylon. Note the warning triangles on the plyons and the painting of the Firestreak , both are for display purposes only. Not sure painting the squadron badge on the tanks ever happened in service, but I could be wrong. Note the fuel filler cap. *Peter Woolley photo.*

Left. I believe this is a combination fuel tank and baggage pod. The two large doors toward the rear of the pod are hinged at the top and provide a lot of storage space. The front seems to have a fuel filler cap, so it seems likely to also be a fuel tank. *Roy Sutherland photo.*

Above. The outer face of the starboard outer light stores pylon. It is mounted just inboard of the wingfold line. Note the various access hatches and stencilling. *Mark Robson photo.*

Right. The outer face of the starboard inner light stores pylon. It is mounted in line with the tail booms. They appear to be identical and may be interchangeable. *Mark Robson photo.*

Left. Looking up at the bottom of the port inboard weapons pylon, showing the pylon cover installed. The stencil warns to keep cover on unless accessing air connections. *Mark Robson photo.*

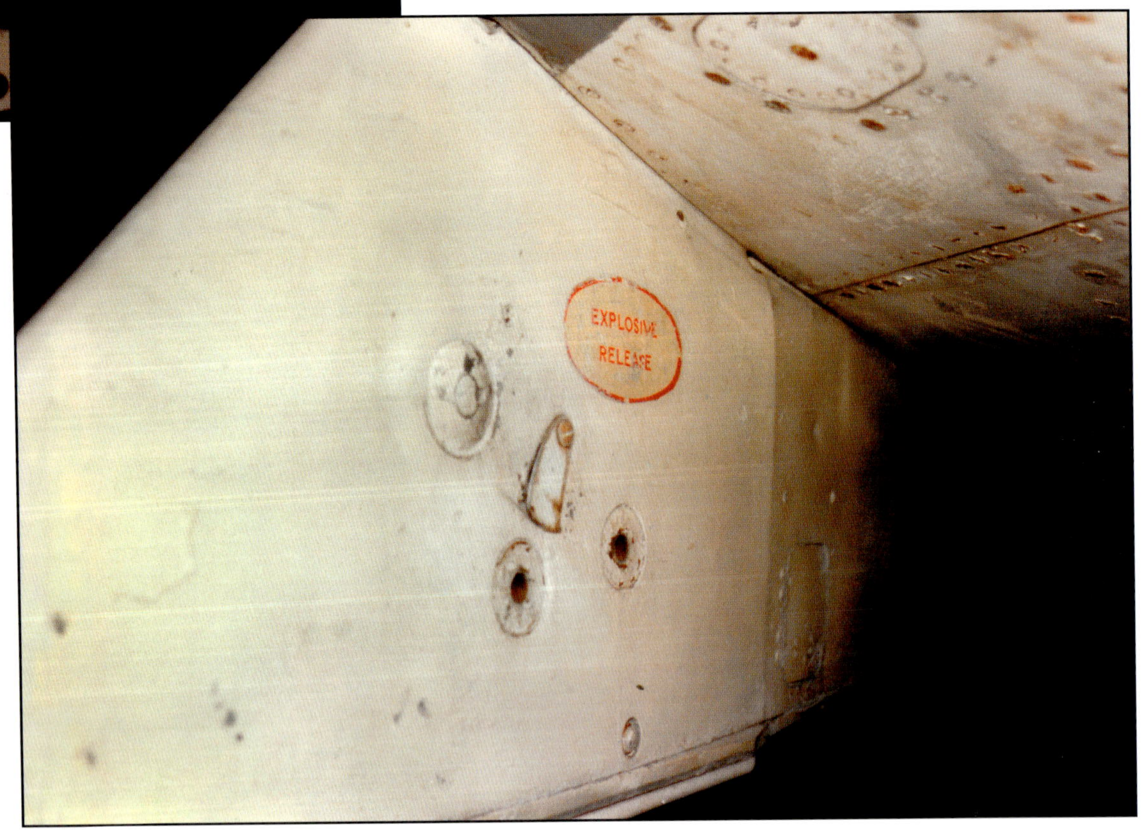

Right. The inner face of the starboard outer light stores pylon. Note the explosive release warning decal, and the holes, which are winch points for lifting the stores into position. *Mark Robson photo.*

Left. The front end of the Bullpup missile. The Bullpup was a US designed and built radio guided air to ground missile, flown by joystick to the target by the pilot. It carried a 250 pound warhead. *John Heck photo.*

Right. The rear end of the Bullpup. With a length of over thirteen and a half feet long, and a weight of over 1700 lbs. the Bullpup was a large missile. A brightly burning tracer on the tail helped the pilot follow the missile in flight. *John Heck photo.*

Top Left. The Red Top missile, attached, with its adaptor, to the inner light stores pylon. The Redtop was a development of the Firestreak AAM, and was carried only by the FAW 2. *Mark Robson photo.*

Facing Bottom Left. Originally dubbed Firestreak Mk. IV, the almost complete redesign led to the missile being named Red Top. On the right is one of a pair of windows for the much improved "Green Garland" infrared proximity fuse. *Mark Robson photo.*

Facing Top Right. A closer look at the pylon adapter from head-on. Manufactured by Hawker-Siddeley, the Red Top had much improved maneuverability and range over its predecessor, but still required dedicated equipment on the aircraft to operate it, unlike the AIM-9 series. *Mark Robson photo.*

Top Left. This Red Top is actually an inert dummy round, blue being the color used for inert weapons. This view shows the enlarged fixed main wings that helped make it much more maneuverable. The annular blast fragmentation warhead was moved forward from the tail and increased in size to 68.3 lbs. *Mark Robson photo.*

Bottom Right. A close-up look at the totally redesigned seeker, named "Violet Banner". It was much more sensitive and was capable of a wider angle of tracking, not having to be limited to tail chase interception. The one piece cast glass cover improved the seekers performance. *Mark Robson photo.*

Top Right. The nozzle for the larger and more powerful "Linnet" rocket motor gave the Red Top a range of 7.5 miles and a top speed of over mach 3 at burnout. The triangular steering fins made it a very agile weapon. *Mark Robson photo.*

Above. The de Havilland Firestreak missile carried by the FAW 1. It is an rocket propelled air-to-air missile that passively homes in on its target's infrared signature. Note the large fins which acted as wings, and the small rectangular fins in the back which were controlled by inputs from the seeker head. *Chris Hughes photo.*

Top Left. Note the twin row of triangular windows that were the proximity fuses, and the 8 sided seeker head. The pylon adapter is seen attached to the missile, this stayed with the aircraft after the missile was fired. The 50 lb. annular blast warhead was mounted in the rear. *Chris Hughes photo.*

Above. The seeker head was cooled by anhydrous ammonia to almost -200°C to improve performance. In flight, the seeker was slaved to the Sea Vixen's Airborne Interception radar system. Once locked on, the missile became a fire-and-forget weapon. *Chris Hughes photo.*

Twin Booms - FAW 1 Tail Booms

Top Left. This image shows the graceful contours of the FAW 1 tailbooms, before the Pintle (or Boom) tanks were added to increase the limited range on the FAW 2. Some prefer this clean look, while others like the more aggressive look of the later variant. *Mark Robson photo.*

Bottom Left. A closer look at the front of the port boom as it fairs into the wing. Surrounding the boom fairing are the heavy gauge doubler plates, for added strength in this crucial area. Ahead of the boom is an access hatch for the mounting of the inner light stores pylon. *Mark Robson photo.*

Top Right. This head-on shot provides a good idea of the cross section of the boom. The doubler plates are very prominent in this view, as is the access hatch. *Mark Robson photo.*

Top Left. A deceptive shot. The canopy in the background is not the Sea Vixen's. This is actually a shot of the inner surface of the starboard tail boom, looking forward from the trailing edge of the inner wing. Limited access to the interior of the forward booms is provided by these two small removable panels. *Mark Robson photo.*

Center Left. The inner face of the port tailboom, showing clearly the few degrees of upward crank that was added to the tailbooms to increase clearance on takeoff. This joint is stiffened internally with a skin doubler plate, as well as this external plate. *Mark Robson photo.*

Bottom Left. The outer face of the port tailboom is clearly seen in this photo. Note the unidentified blade antenna on top of the forward portion of the boom. Please contact us if you know its purpose. *Mark Robson photo.*

Top Right. Close-up image of the NACA duct that feeds cooling air to the electronics bay in the port tailboom. Also shown to good effect are the many screws that attach the rear tailboom assemblies to the wing. An internal stiffoning ring adds further strength. *Mark Robson photo.*

Twin Booms - FAW 2 Tail Booms

Top. A head-on view of the boom tanks that are the most visible identification point of the Sea Vixen FAW 2. The complex system of fairing panels that blend the overwing tanks in is evident. *Mark Robson photo.*

Top Right. While it appears that the boom tanks are painted a darker color, it is a trick of the light caused by the change in angle from the original boom to the scabbed on tank. Note the 3 blade aerials on top of the stbd. boom. Purpose unknown. Constant upgrades and retrofits are still common on military equipment . *Roy Sutherland photo.*

Bottom Right. A view looking forward along the inside face of the port tailboom, showing the stepped nature of the boom tanks installed. Note the rubber sealing strip that runs around the perimeter, the additional access hatches, and the offset mounted blade antenna . *Roy Sutherland photo.*

Top. This image of the outer face of the starboard tailboom shows two access panels. The forward one is an electrical connection port, and the aft on is for charging the two hydraulic accumulator tanks. Stbd boom is similarly fitted. *Chris Hughes photo.*

Top Right. A nice shot showing the overwing portion on the port boom tank. It rained lightly the entire day we were shooting and measuring the aircraft. Mike and I froze our butts off on this chilly November day, but it was worth it! *Roy Sutherland photo.*

Bottom Right. Lovely weather! Mike holds the scale, which is in inches and feet. Leave it to de Havilland to make even these bolt on tanks attractive in a purposeful way. Note the 4 prominent access hatches on the nose of each boom tank. *Roy Sutherland photo.*

Top Left. View looking forward and outboard of the top of the port vertical tail and slab tailplane, showing to advantage the upper aerodynamic fairing between the two. *Mark Robson photo.*

Top Right. Looking down on the starboard vertical tailplane. The keyhole slot in the trailing edge of the elevator is just visible. There are four keyhole slots, one aft of each elevator hinge. The elevator is linked to the flap system, and deflects upwards when the flaps are deployed to counter the trim changes they create. *Roy Sutherland photo*

Bottom Left. Looking forward across the top of the slab tailplane. This incorporates both an elevator, and an all flying adjustable tailplane. The elevator is, in practice, a giant trim tab, and the control of the all flying tailplane is geared for different speeds and stick loadings. *Mark Robson photo..*

Top Left. This view shows the elevator deflected upward, showing the mass balances that tuck into slots in the underside of the slab tailplane. The rudder controls are duplicated, in that the port control run also crosses through the stab to the stbd. rudder. *Roy Sutherland photo.*

Bottom Left. Inside of port vertical tailplane showing the lower fin to stab fairing. The elevator is not capable of downward deflection, as the mass balances cannot pass through the upper surface of the slab tailplane. The upper surface color wraps around under the leading edge of the stab. *Roy Sutherland photo.*

Facing Top Left. Only the port tailplane fairing is fitted with a pair of formation lights. The back of the prototype's tail fairings were chopped off to provide clearance for tight hangar deck and deck elevator clearances. *Mark Robson photo.*

Facing Top Right. A clear shot of the outer face of port tailplane of Sea Vixen FAW2 serial XJ565. The large rudder area and fixed trim tabs are noteworthy. *Roy Sutherland photo.*

Facing Bottom Right. The outer face of the starboard tailplane reveals many details in this shot. Note the numerous, oddly shaped access panels, the cutout for the oval rudder mass balance, and the hydraulic tail bumper. *Chris Hughes photo.*

Twin Booms - Tailplanes continued

Left. Looking out at the port inner face of the vertical tailplane. A field of access panels and stiffening plates is apparent. My longtime friend Mike Grzebien helps with measurements on a less than stellar weather day at the Newark Air Museum. Inset images show both the rudder deflected showing its odd oval mass balances, and the damped tail bumpers. *Roy Sutherland* photo.

Bottom Right. This illustration is terrific as a tool to understand how the Sea Vixen's tailplanes were put together. FAW 1 and FAW 2 tailplanes were virtually identical. Only the pintle tanks of the latter, which were added on to the standard Mk 1 booms, set them apart. © *Crown Copyright.*

Right. The surprisingly complex internal structure of the Sea Vixen's tail booms. Note the pivot point and the rails for the all flying horizontal tailplane. It would be a safe bet to speculate that all the reinforcing plates were added due to the need to land this beast on carriers. © *Crown Copyright.*

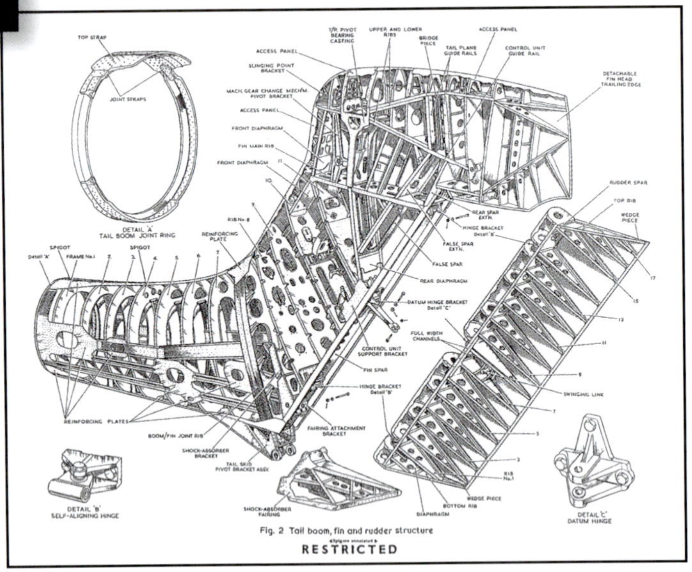

Fig. 2 Tail boom, fin and rudder structure

RESTRICTED

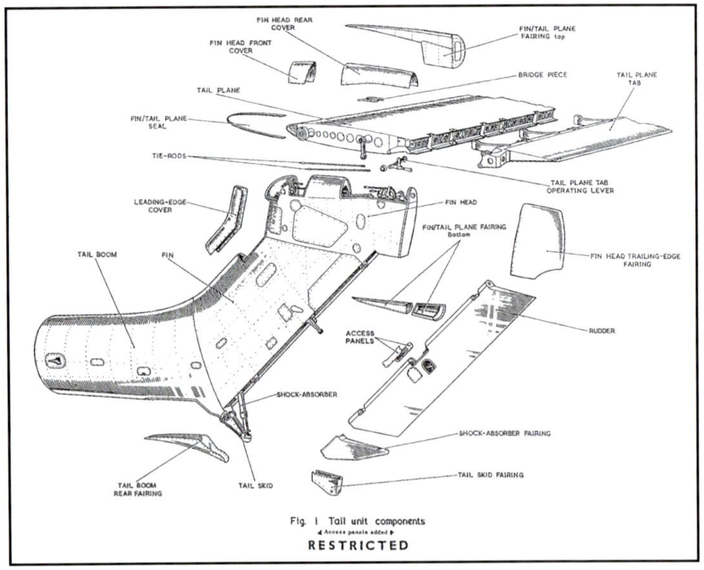

Fig. 1 Tail unit components

RESTRICTED

Sea Vixen D.3 XP924

Top Left. XP924 was one of three Sea Vixen's converted to a D.3 pilotless drone after retirement. Seen here in drone colors, it flew for years from RAE Llanbedr. *Keith Bunyan photo.*

Top Right. Sea Vixen XP924 was built as an FAW.2 and flew with 899 Squadron. The last surviving drone and the last flying Sea Vixen, this aircraft still flies today. *Peter Steehouwer photo.*

Bottom Left. After the Drone program ended XP924 was sold into private hands, as the only remaining airworthy Sea Vixen, and registered as G-CVIX. *Robin Powell photo.*

Top Left. A very expensive aircraft to keep flying, A very generous sponsor was found when Red Bull © agreed to fund XP924 for the 2004-2007 airshow seasons. *Robin Powell photo.*

Top Right. Painted up in a very bold and colorful Red Bull© scheme, the Sea Vixen became a well loved performer at airshows in the UK and Europe. *Robin Powell photo.*

Bottom Left. While hardcore aviation enthusiasts complained about the non-historical color scheme, no one complained about having a flying Sea Vixen! *Peter Steehouwer photo.*

Sea Vixen D.3 XP924 continued

Top Left. The Red Bull© color scheme was very popular with airshow crowds. Unfortunately, the sponsorship came to an end in 2007, and the future is uncertain. *Peter Steehouwer photo.*

Bottom Left. Funding to keep G-CVIX flying was found for the 2008 season, and XP924 flew a few shows in the UK. *Roland Bogush photo.*

Bottom Right. Repainted in her original 899 Squadron livery, aviation enthusiasts were thrilled to see this All Weather Fighter flying again in “proper” markings. *Roland Bogush photo.*

Top Left. G-CVIX continued to fly, performing at a handful of venues during the 2009 airshow season. *Roland Bogush photo.*

Bottom Left. de Havilland Aviation, with additional funding from the Royal Navy, are keeping XP924 flying. The 2010 season is, if you will forgive the pun, still up in the air. *Roland Bogush photo.*

Bottom Right. The Sea Vixen is truly unique amongst airshow aircraft; a twin boom, twin engine Royal Navy jet fighter that flew during the 1960s and 70s. Amazing. *Roland Bogush photo.*

Sea Vixen D.3 XP924 continued

Flightline Series 1
de Havilland Sea Vixen FAW 1 and 2 British Naval All Weather Jet Fighter
An in-depth photographic study by Roy Sutherland

Flightline Series 1: de Havilland Sea Vixen FAW 1 and 2 end.

BarracudaGraphs **Flightline Series 1**

Langhorne Publishing House, Inc.

de Havilland Sea Vixen FAW 1 and 2

British Naval All Weather Jet Fighter